INTRODUCTION TO

Gauge Integrals

INTRODUCTION TO

Gauge Integrals

Charles Swartz
Department of Mathematical Sciences
New Mexico State University, USA

World Scientific
Singapore • New Jersey • London • Hong Kong

Published by

World Scientific Publishing Co. Pte. Ltd.

P O Box 128, Farrer Road, Singapore 912805

USA office: Suite 1B, 1060 Main Street, River Edge, NJ 07661

UK office: 57 Shelton Street, Covent Garden, London WC2H 9HE

Library of Congress Cataloging-in-Publication Data

Swartz, Charles, 1938–

Introduction to gauge integrals / Charles Swartz.

p. cm.

Includes bibliographical references and index.

ISBN 9810242395 (alk. paper)

1. Gauge integrals. I. Title.

QA311 .S88 2001

515'.43--dc21 00-068521

British Library Cataloguing-in-Publication Data

A catalogue record for this book is available from the British Library.

This book is printed on acid-free paper.

Printed in Singapore by Uto-Print

Preface

Although it is well-recognized that the Lebesgue integral with its superior convergence properties is superior to the Riemann integral, most introductory real analysis texts contain an exposition of the Riemann integral because it is much easier to describe than the Lebesgue integral. However, there is an integral, called the gauge (Henstock-Kurzweil, Riemann complete or generalized Riemann) integral, which is but a slight variation of the Riemann integral, which possesses all the convergence properties of the Lebesgue integral, is actually more general than the Lebesgue integral and is no more difficult to describe than the Riemann integral. Despite its ease of exposition and its powerful convergence properties, the gauge integral has not found its way into very many introductory real analysis texts (see, however, [DS], [M]). This may be due to the fact that many people are not yet familiar with the gauge integral and its properties (see [Ba1], for example), or it may be due to simple inertia. This text contains an elementary exposition of the gauge integral. The text can be read by someone simply wanting to become acquainted with the gauge integral, or there is a sufficient supply of exercises so that the material on the gauge integral could be incorporated into a real analysis course.

There is a variant of the gauge integral introduced by McShane, called the McShane integral, which is actually equivalent to the Lebesgue integral over Euclidean spaces, but, as a gauge-type integral, has a much simpler exposition than the usual definitions of the Lebesgue integral. This integral is also not very well-known but is a good candidate for inclusion in an introductory real analysis course. We also give an exposition of the McShane integral. As with the gauge integral, this material could also be used as a supplement for a real analysis course.

The McShane integral is an absolute integral whereas the gauge integral is a conditional integral which is more general than the McShane (Lebesgue) integral. We give an elementary proof due to R. Vyborny of the fact that a function is McShane integral if and only if it is absolutely gauge integrable thus establishing the relationship between the two integrals (Chapter 10).

This text is an elementary introduction to gauge-type integrals and, except for the starred sections, requires only basic concepts encountered in an introductory real analysis course. The text is written in such a way that an instructor of an introductory real analysis course can easily use the material in the text to give an exposition of either the gauge or McShane integral. For the benefit of readers who are familiar with the Lebesgue integral, there are a few more advanced topics discussed in the text which require a stronger background; these sections and theorems are indicated by asterisks and are not prerequisites for other sections in the text. For readers who are interested in studying more advanced properties of the gauge-type integrals, we give statements and a few proofs of advanced results and also references in the Notes/Remarks section following each chapter and in Appendices 3 and 4.

In the first chapter we use the Fundamental Theorem of Calculus (FTC) to motivate the definition of the gauge integral over bounded intervals and ultimately show that the FTC holds in full generality for the gauge integral. The basic properties of the integral, such as linearity, positivity, etc., are developed in Chapter 2. All of the proofs of the major convergence theorems for the gauge (and McShane) integral are based on a result called Henstock's Lemma; this result is proven in Chapter 3. Also, in Chapter 3 Henstock's Lemma is used to show that there are no improper integrals (over bounded intervals) for the gauge integral. It is very easy to extend the definition of the gauge integral to unbounded intervals in the real line; this extension is carried out in Chapter 4. The major convergence theorems, the Monotone and Dominated Convergence Theorems, are proven in Chapter 5. There are also several applications of the convergence theorems given in Chapter 5.

Chapters 6 and 7 are starred chapters which contain more advanced material and can be skipped by readers who are interested in learning only the basic facts about the gauge integral. In Chapter 6 the gauge integral is used to define and develop some of the basic properties of Lebesgue measure on the real line. The results are used to give more general forms of the Monotone and Dominated Convergence Theorems. The space of gauge integrable functions carries a natural (semi) norm, called the Alexiewicz norm, and Chapter 7 contains a study of this function space including references to some of the more technical results pertaining to this space.

In Chapter 8 the gauge integral is extended to n-dimensional Euclidean space. We present two versions of Fubini's Theorem which relate multiple integrals to iterated integrals. The first version restricts the class of functions considered but is general enough to cover applications typically encountered in beginning analysis courses. The second (starred) version of the Fubini Theorem is quite general but requires some of the results from Chapter 6 on Lebesgue measure.

Chapter 9 contains an exposition of the McShane integral. The McShane integral is obtained from a "slight" but very important variant of the gauge integral; however, this change has a very profound effect on the resulting integration theory leading to an absolute integral which is equivalent to the Lebesgue integral (Appendix 4). Many of the proofs of the basic properties of the McShane integral are virtually identical to the proofs of the corresponding properties of the gauge integral so only references to the gauge integral proofs are given.

Chapter 10 contains an elementary proof due to R. Vyborny that a function is McShane integrable if and only if it is absolutely gauge integrable. The Notes/Remarks section of this chapter contains references to other proofs of this result which use more advanced techniques.

Chapters 1-5 plus the unstarred sections in Chapter 8 should provide an exposition of the gauge integral suitable for inclusion in an introductory real analysis course. Similarly, Chapter 9 (along with the references to earlier gauge integral proofs) should serve as an exposition of the McShane integral which is suitable for an introductory real analysis course. There should be a sufficient supply of exercises in either choice.

I would like to thank Rich Reynolds for reading through most of the manuscript and making many suggestions and corrections. I would also like to thank Rose Marquez for doing her usual exemplary job of typing the manuscript.

We give a list of some of the notation which will be employed in the text.

$\mathbb{N}$ positive integers

$\mathbb{Q}$ rational numbers

$\mathbb{R}$ real numbers

$\mathbb{C}$ complex numbers

C_E characteristic function of a set E

$\uparrow$ increasing sequence or function

$\downarrow$ decreasing sequence or function

$\bigvee$ supremum

$\bigwedge$ infimum

$\underline{\lim}$ limit inferior

$\overline{\lim}$ limit superior

Contents

Chapter 1

Introduction to the Gauge or Henstock-Kurzweil Integral

It is well-recognized that the Lebesgue integral is much superior to the Riemann integral because it integrates more functions and possesses convergence theorems which are much more general than those which are valid for the Riemann integral. However, the Lebesgue integral is technically much more difficult to describe than the Riemann integral, and for this reason most introductory real analysis texts contain presentations of the Riemann integral or its immediate relative the Riemann-Stieltjes integral rather than the superior Lebesgue integral ([Ru], [Ho]). However, there is an integral, called the gauge or Henstock-Kurzweil integral, which is essentially as easy to describe as the Riemann integral and which possesses all the advantages of the Lebesgue integral and even more.

In this introductory chapter we give the definition of the gauge or Henstock-Kurzweil integral and give several examples which illustrate the generality of the gauge integral. The definition is but a slight variation of the definition of the classical Riemann integral (see Appendix 1 for the Riemann integral), but the effect of the change is very profound, yielding an integral more general than the Lebesgue integral, having no "improper integrals", satisfying very general convergence properties like those of the Lebesgue integral and for which the Fundamental Theorem of Calculus holds in full generality.

We motivate the definition of the gauge integral by considering the Fundamental Theorem of Calculus (FTC). If $f : [a, b] \to \mathbb{R}$ is a function which has a derivative f' everywhere in $[a, b]$, then the desired conclusion of the Fundamental Theorem is

$$\int_a^b f' = f(b) - f(a)\,.$$

Volterra constructed an example of a function $f : [0,1] \to \mathbb{R}$ with a bounded derivative f' on $[0,1]$ which is not Riemann integrable; that is, the FTC does not hold in full generality for the Riemann integral (see [BBT] § 1.18 and Exercise 5.5, [Sw1] § 3.3). The version of the FTC for the Riemann integral requires the assumption that the derivative f' is Riemann integrable ([Ru] 6.16); the Lebesgue integral suffers the same "defect" (see [N1] § IX.7, [Sw1] § 4.3). However, suppose that we have a differentiable function $f : [a,b] \to \mathbb{R}$, and we consider how we might attempt a proof of the Fundamental Theorem for the Riemann integral. Let $P = \{a = x_0 < x_1 \ldots < x_n = b\}$ be a partition of the interval $[a,b]$. In each subinterval $[x_i, x_{i+1}]$ we seek a point t_i such that the term $f'(t_i)(x_{i+1} - x_i)$ approximates $f(x_{i+1}) - f(x_i)$, the desired value of the integral of f' over the interval $[x_i, x_{i+1}]$ via the formula in the Fundamental Theorem. If this choice of points is possible in each subinterval, then the Riemann sum $\sum_{i=0}^{n-1} f'(t_i)(x_{i+1} - x_i)$ should furnish an approximation to the desired value of the integral of f' over $[a,b]$ since $\sum_{i=0}^{n-1}(f(x_{i+1}) - f(x_i)) = f(b) - f(a)$. Thus, the first question we might ask is whether such a choice of points is always possible. The following lemma shows that in some sense this is the case.

Lemma 1 (*Straddle Lemma*). *Let $f : [a,b] \to \mathbb{R}$ be differentiable at $z \in [a,b]$. Then for each $\varepsilon > 0$ there is a $\delta > 0$ such that*

$$|f(v) - f(u) - f'(z)(v-u)| \le \varepsilon(v-u)$$

whenever $u \le z \le v$ and $[u,v] \subset [a,b] \cap (z-\delta, z+\delta)$.

Proof: Since f is differentiable at z, there is a $\delta > 0$ such that

$$\left|\frac{f(x) - f(z)}{x - z} - f'(z)\right| < \varepsilon$$

for $0 < |x - z| < \delta$, $x \in [a,b]$. If $z = u$ or $z = v$, then the conclusion of the lemma is immediate so suppose $u < z < v$. Then

$$\begin{aligned} |f(v) - f(u) - f'(z)(v-u)| &\le |f(v) - f(z) - f'(z)(v-z)| \\ &\quad + |f(z) - f(u) - f'(z)(z-u)| \\ &< \varepsilon(v-z) + \varepsilon(z-u) = \varepsilon(v-u)\,. \end{aligned}$$

The geometrical interpretation of the Straddle Lemma is clear. If the points u and v "straddle" z, then the slope of the chord between the points $(u, f(u))$ and $(v, f(v))$ is close to the slope of the tangent line at $(z, f(z))$. The conclusion of the lemma fails if the points u and v do not straddle z; consider $f(t) = t^2 \cos(\pi/t)$ for $t \neq 0$ and $f(0) = 0$ near $t = 0$.

If it is now possible to partition the interval $[a, b]$ in such a way that the point $t_i \in [x_i, x_{i+1}]$ and the subinterval $[x_i, x_{i+1}]$ satisfy the conclusion of the Straddle Lemma, we obtain

$$(1) \qquad \left| \sum_{i=0}^{n-1} f'(t_i)(x_{i+1} - x_i) - [f(b) - f(a)] \right|$$

$$= \left| \sum_{i=0}^{n-1} \{ f'(t_i)(x_{i+1} - x_i) - [f(x_{i+1}) - f(x_i)] \} \right|$$

$$< \varepsilon \sum_{i=0}^{n-1} (x_{i+1} - x_i) = \varepsilon(b - a)$$

so the Riemann sum $\sum_{i=0}^{n-1} f'(t_i)(x_{i+1} - x_i)$ give a good approximation to the desired value of the integral of f' over $[a, b]$, namely, $f(b) - f(a)$.

At this point we consider the (Riemann) integrability of the function f' over the interval $[a, b]$. Recall that a function $g : [a, b] \to \mathbb{R}$ is Riemann integrable over $[a, b]$ if there exists $A \in \mathbb{R}$ such that for every $\varepsilon > 0$ there exists a $\delta > 0$ such that if $P = \{a = x_0 < x_1 < \cdots < x_n = b\}$ is a partition of $[a, b]$ with

$$(2) \quad \max\{x_{i+1} - x_i : 0 \leq i \leq n - 1\} < \delta \text{ and } t_i \in [x_i, x_{i+1}],$$

then $|\sum_{i=0}^{n-1} g(t_i)(x_{i+1} - x_i) - A| < \varepsilon$. Of course, A is the Riemann integral of g over $[a, b]$ and is denoted by $\int_a^b g$. [See Appendix 1.]

Note that the $\delta = \delta(z)$ given in the conclusion of the Straddle Lemma depends on the point z in the interval $[a, b]$ and will undoubtedly vary as z runs through $[a, b]$ depending upon the behavior of the function f. That is, there may not exist a single positive constant δ which will satisfy the conclusion of the Straddle Lemma for all points of the interval $[a, b]$, and, thus, we cannot guarantee the Riemann integrability of the derivative f'. It is exactly this requirement of being able to uniformly partition the interval and not take into account the local behavior of the function being integrated which so severely restricts the applicability of the Riemann integral.

However, suppose we rewrite condition (2) in the following form:

(3) $t_i - \delta < x_i \leq t_i \leq x_{i+1} < t_i + \delta$.

Our difficulty is that we cannot guarantee the existence of a positive constant δ which will satisfy (3) and the conclusion of the Straddle Lemma for all $t_i \in [a, b]$. Since for every $t \in [a, b]$ we can find a $\delta(t) > 0$ satisfying the conclusion of the Straddle Lemma, this suggests that we consider replacing the constant δ in (3) by $\delta(t_i)$, i.e., we replace (3) by

(4) $t_i - \delta(t_i) < x_i \leq t_i \leq x_{i+1} < t_i + \delta(t_i)$.

We can then guarantee that the Straddle Lemma applies across the entire interval $[a, b]$ and the calculation in (1) shows that the Riemann sum for any partition and choice of points satisfying (4) will give a good approximation to the desired value of the integral. Thus, we should consider defining a "new" integral by simply replacing the constant δ in the definition of the Riemann integral by a positive function $\delta : [a, b] \to (0, \infty)$. We refer to the resulting integral as the *gauge integral* (this integral is also called the *Henstock-Kurzweil integral* or the *Riemann complete integral* or the *generalized Riemann integral*). We next give a formal definition of the gauge integral.

A *tagged partition* of an interval $I = [a, b]$ is a finite set of ordered pairs $\mathcal{D} = \{(t_i, I_i) : 1 \leq i \leq m\}$ where $\{I_i : 1 \leq i \leq m\}$ is a partition of I consisting of closed non-overlapping subintervals and t_i is a point belonging to I_i; t_i is called the *tag* associated with I_i. If $f : I \to \mathbb{R}$, the *Riemann sum* of f with respect to $\mathcal{D}$ is defined to be

$$S(f, \mathcal{D}) = \sum_{i=1}^{m} f(t_i)\ell(I_i),$$

where $\ell(I_i)$ is the length of the subinterval I_i. It is convenient for future use when we consider integration over unbounded intervals to rewrite (4) in another form. If $\delta : I \to (0, \infty)$ is a positive function, we define an open interval valued function on I by setting

$$\gamma(t) = (t - \delta(t), t + \delta(t)).$$

If $I_i = [x_i, x_{i+1}]$, we can rewrite (4) in the form

(5) $t_i \in I_i \subset \gamma(t_i)$.

Any function γ defined on I such that $\gamma(t)$ is an open interval containing t for each $t \in I$ is called a *gauge* on I. If $\mathcal{D} = \{(t_i, I_i) : 1 \le i \le m\}$ is a tagged partition of I and γ is a gauge on I, we say that $\mathcal{D}$ is *γ-fine* if (5) is satisfied; if $\mathcal{D}$ is γ-fine, we write $\mathcal{D} << \gamma$. We now have the following generalization of the classical Riemann integral.

Definition 2. *Let $f : [a,b] \to \mathbb{R}$. f is said to be (gauge) integrable over $[a,b]$ if there exists $A \in \mathbb{R}$ such that for every $\varepsilon > 0$ there exists a gauge γ on $[a,b]$ such that $|S(f,\mathcal{D}) - A| < \varepsilon$ whenever $\mathcal{D}$ is a γ-fine tagged partition of $[a,b]$.*

The number A is called the (gauge) integral of f over $I = [a,b]$ and is denoted by $\int_a^b f$ or $\int_I f$; when we encounter integrals depending upon parameters, it is also convenient to write $\int_a^b f(t)dt$ or $\int_I f(t)dt$ (see Theorem 4 below). In order for Definition 2 to make sense, there is one matter that needs to be addressed. We must show that every gauge γ has at least one γ-fine tagged partition.

Theorem 3. *Let γ be a gauge on the interval $I = [a,b]$. There exists a γ-fine tagged partition $\mathcal{D}$ of I.*

Proof: Define a subset E of I by $E = \{t \in I : \exists.$ a tagged partition of $[a,t]$ which is γ-fine$\}$. First, $E \neq \emptyset$, for we can choose $x \in \gamma(a)$ such that $a < x < b$. Then $\{(a,[a,x])\}$ is a γ-fine tagged partition of $[a,x]$. Let $y = \sup E$. We claim $y \in E$. Choose $x \in E$ such that $x \in \gamma(y)$ and $x < y$. There is a γ-fine tagged partition $\mathcal{D}$ of $[a,x]$. The set $\mathcal{D} \cup \{(y,[x,y])\}$ is a γ-fine tagged partition of $[a,y]$, i.e., $y \in E$.

Next, we claim that $y = b$. Suppose $y < b$. Choose $w \in \gamma(y) \cap (y,b)$. Let $\mathcal{D}$ be a γ-fine tagged partition of $[a,y]$. Then $\mathcal{D}' = \mathcal{D} \cup \{(y,[y,w])\}$ is a γ-fine tagged partition of $[a,w]$. This contradicts the definition of y. Thus, $y = b$ as desired.

See also Exercise 4 where another method of proof is outlined.

We also need to address the uniqueness of the integral, justifying the notation employed above.

Theorem 4. *The number A in Definition 2 is unique.*

Proof: Let $\varepsilon > 0$. Suppose A_1 and A_2 satisfy the conditions in Definition 2 with respect to the gauges γ_1 and γ_2, respectively. Let $\gamma(t) = \gamma_1(t) \cap \gamma_2(t)$. By Exercise 1, if $\mathcal{D}$ is a γ-fine tagged partition, then $\mathcal{D} << \gamma_i$ for $i = 1, 2$, so

$$|A_1 - A_2| \le |A_1 - S(f,\mathcal{D})| + |S(f,\mathcal{D}) - A_2| < 2\varepsilon\,.$$

Thus, $A_1 = A_2$.

From the remarks above we see that a (bounded) function is Riemann integrable if and only if it is gauge integrable with respect to gauges whose values are open intervals which are symmetric about the points in the domain of the gauge and have constant length.

From our previous discussion above we can easily establish the Fundamental Theorem of Calculus (FTC) in full generality.

Theorem 5 (*FTC*). *Suppose $f : [a, b] \to \mathbb{R}$ is differentiable at every point of $[a, b]$. Then f' is integrable over $[a, b]$ and*

$$\int_a^b f' = f(b) - f(a).$$

Proof: Let $\varepsilon > 0$. For each $t \in [a, b]$ let $\delta(t)$ be given by the Straddle Lemma and set $\gamma(t) = (t - \delta(t), t + \delta(t))$. If $\mathcal{D} = \{(t_i, I_i) : 1 \le i \le m\} << \gamma$, then the calculation in (1) shows that f' is integrable over $[a, b]$ with $\int_a^b f' = f(b) - f(a)$.

The important point in Theorem 5 is that any arbitrary derivative is (gauge) integrable. The versions of the FTC for both the Riemann and the Lebesgue integrals require the assumption that the derivative is either Riemann or Lebesgue integrable, respectively [see [Ru] 6.16, [Sw1] 4.3.3].

Exercise 10 gives a generalization of Theorem 5.

It follows from Theorem 5 that the rules developed in elementary calculus for calculating integrals using anti-derivatives hold for the gauge integral. Another common method used in calculating integrals is the method of substitution or change of variable. We may use Theorem 5 to establish such a result for the gauge integral. In what follows if $f : [a, b] \to \mathbb{R}$, $a < b$, we set $\int_b^a f = -\int_a^b f$ and $\int_a^a f = 0$.

Theorem 6 (*Integration by Substitution*). *Let $f : [a, b] \to \mathbb{R}$ and $\varphi : [\alpha, \beta] \to [a, b]$ be differentiable functions. Then*

$$\int_{\varphi(\alpha)}^{\varphi(\beta)} f' = \int_a^b (f' \circ \varphi)\varphi'.$$

Proof: By the Chain Rule $(f \circ \varphi)' = (f' \circ \varphi)\varphi'$. Theorem 5 gives $\int \alpha\beta(f' \circ \varphi)\varphi' = f(\varphi(\beta)) - f(\varphi(\alpha)) = \int_{\varphi(\alpha)}^{\varphi(\beta)} f'$.

We now give two examples which like the proof of the FTC illustrate the advantage and generality achieved by allowing the intervals in the gauge to have variable lengths.

Example 7. *Suppose $f : [a,b] \to \mathbb{R}$ has the constant value c except at a countable number of points $E = \{z_k : k \in \mathbb{N}\}$. We show that f is integrable over $[a,b]$ with $\int_a^b f = c(b-a)$.*

Let $\varepsilon > 0$. If $\mathcal{D} = \{(t_i, I_i) : 1 \le i \le n\}$ is a tagged partition of $[a,b]$, consider

$$|S(f,\mathcal{D}) - c(b-a)| = \left| \sum_{i=1}^{n} \{f(t_i) - c\}\ell(I_i) \right| . \tag{6}$$

If $t_i \notin E$, the term $(f(t_i) - c)\ell(I_i)$ in (6) is 0 so it is reasonable to start defining a gauge by setting $\gamma(t) = (t-1, t+1)$ if $t \notin E$. If $t_i = z_k$ for some k and if $\mathcal{D} << \gamma$ for some gauge γ, then $|f(t_i) - c|\ell(I_i) \le |f(z_k) - c|\ell(\gamma(z_k))$. So if we choose $\delta_k = \varepsilon/(|f(z_k) - c|)2^{k+2}$ and set $\gamma(z_k) = (z_k - \delta_k, z_k + \delta_k)$, then when $\mathcal{D} << \gamma$ and $t_i = z_k$ we have $|f(t_i) - c|\ell(I_i) < \varepsilon/2^{k+1}$. If $\mathcal{D} << \gamma$, we have from (6) that

$$|S(f,\mathcal{D}) - c(b-a)| < 2\sum_{k=1}^{\infty} \varepsilon/2^{k+1} = \varepsilon$$

since each z_k can be the tag for at most 2 subintervals in $\mathcal{D}$.

In particular, if $f(t) = 0$ for t irrational and $f(t) = 1$ for t rational (the *Dirichlet function*), then f is integrable over any interval $[a,b]$ with $\int_a^b f = 0$. The Dirichlet function is the usual example that is given for a (bounded) function which is not Riemann integrable. The computation above in Example 7 illustrates the advantage of using gauges with variable length. [For a more general result see Example 10.]

We next consider an example of an unbounded function which has an improper Riemann (or Cauchy-Riemann) integral.

Example 8. *Let $f(x) = 1/\sqrt{x}$ for $x > 0$ and $f(0) = 0$. We claim that f is integrable over $[0,1]$ with $\int_0^1 f = 2$ as in the calculus. Let $\varepsilon > 0$. First consider the function near 0. If $0 < x < 1$, from the calculus the area under the curve over $[0,x]$ is $2\sqrt{x}$. If we construct the gauge γ so that $\gamma(t) \subset (0,t)$ for every $t \in (0,1]$, then when $\mathcal{D}$ is a γ-fine tagged partition of $[0,1]$ the tag associated with the subinterval of $\mathcal{D}$ containing 0 must be 0 [Exercise 8]. If we set $\gamma(0) = (-\varepsilon^2/16, \varepsilon^2/16)$, then*

$$|f(0)(x_1 - 0) - 2\sqrt{x_1}| = 2\sqrt{x_1} < 2\sqrt{\epsilon^2/16} = \varepsilon/2$$

whenever $[0, x_1] \subset \gamma(0)$. *If* $0 < u < v \le 1$, *the area under the curve over* $[u, v]$ *is* $2\sqrt{v} - 2\sqrt{u}$ *and if* $u \le z \le v$, *we have*

$$\begin{aligned} |f(z)(v-u) - (2\sqrt{v} - 2\sqrt{u})| &= (v-u)\left|\frac{1}{\sqrt{z}} - \frac{2}{\sqrt{v}+\sqrt{u}}\right| \\ &= \frac{v-u}{z}\left|\frac{\sqrt{z}(\sqrt{v}+\sqrt{u}) - 2z}{\sqrt{v}+\sqrt{u}}\right| \\ &\le \frac{v-u}{z}\left|\frac{\sqrt{z}(\sqrt{v}+\sqrt{u}) - 2z}{\sqrt{z}}\right| \\ &= \frac{v-u}{z}|\sqrt{v}+\sqrt{u}-2\sqrt{z}| \\ &\le \frac{v-u}{z}(|\sqrt{v}-\sqrt{z}| + |\sqrt{u}-\sqrt{z}|) \\ &= \frac{v-u}{z}\left(\frac{v-z}{\sqrt{v}+\sqrt{z}} + \frac{z-u}{\sqrt{u}+\sqrt{z}}\right) \\ &\le \frac{v-u}{z}\left(\frac{v-z}{\sqrt{z}} + \frac{z-u}{\sqrt{z}}\right) = \frac{(v-u)^2}{z^{2/3}}\,. \end{aligned}$$

This suggests that we define $\delta(z) = \varepsilon z^{3/2}/4$ *and set* $\gamma(z) = (z - \delta(z),\, z + \delta(z)) \cap (0, 2)$ *for* $0 < z$. *Now suppose* $\mathcal{D} = \{(t_i, I_i) : 0 \le i \le n\} << \gamma$, *with* $I_i = [x_i, x_{i+1}], 0 = x_0 < x_1 < \cdots < x_{n+1} = 1$.

Then $t_0 = 0$ *and from the estimate above, we have*

$$\begin{aligned} |S(f, \mathcal{D}) - 2| &= \left|\sum_{i=0}^{n}\{f(t_i)(x_{i+1} - x_i) - 2(\sqrt{x_{i+1}} - \sqrt{x_i})\}\right| \\ &\le 2\sqrt{x_1} + \sum_{i=1}^{n}|f(t_i)(x_{i+1} - x_i) - 2(\sqrt{x_{i+1}} - \sqrt{x_i})| \\ &< \varepsilon/2 + \sum_{i=1}^{n}\frac{(x_{i+1} - x_i)^2}{t_i^{3/2}} \le \varepsilon/2 + \sum_{i=1}^{n}\frac{(x_{i+1} - x_i)}{t_i^{3/2}}2\delta(t_i) \\ &\le \varepsilon/2 + \sum_{i=1}^{n}\frac{\varepsilon(x_{i+1} - x_i)}{2} \\ &= \varepsilon\,. \end{aligned}$$

Note that near the point 0 where the function f has a singularity the gauge is much "finer" than it is near the point 1 when the function is relatively "flat." That is, by using variable length gauges we are able to take into account the local behavior of the function; the Riemann integral is not able to take this local behavior into account. As noted above the function f has an improper or Cauchy-Riemann integral. We will show later that any function which has an improper or Cauchy-Riemann integral is gauge integrable (Theorem 3.4); that is, there are no "improper integrable" functions for the gauge integral (see also Theorem 4.4 for the case of integrals over unbounded intervals).

Notes/Remarks

The gauge integral was originally introduced by J. Kurzweil and used for studies in ordinary differential equations ([K1]). The integral was discovered independently by R. Henstock who developed the major convergence theorems (Monotone and Dominated Convergence Theorems) for the gauge integral ([H1]; see [H2] for an exposition of the integral by Henstock). For texts containing expositions of the gauge integral, see [DS], [Go1], [K2], [LPY], [M], [ML] and [Pf1].

The version of the FTC for both the Riemann and Lebesgue integrals require the assumption that the derivative be either Riemann or Lebesgue integrable, respectively. [See Example 2.12 and [Sw1] § 3.3 and 4.3.1.] This led mathematicians to search for an integration theory for which the FTC holds in full generality. In the 1910's Perron and Denjoy developed theories of integration which extended the Lebesgue theory and for which the FTC holds in full generality as in Theorem 5. These integration theories were later proven to be equivalent and are equivalent to the gauge integral. For an exposition of these integrals and proofs of their equivalence, see [Go1]; for a historical discussion of the Perron and Denjoy integrals, see [Pe].

Exercise 10 gives a generalization of the FTC in Theorem 5 (see also [ST]). Other versions of the FTC for the gauge integral are given in [Sw3], and for an interesting version involving parametric derivatives, see [LA2].

Theorem 3 on the existence of γ-fine tagged partition is often attributed to Cousin ([Co]). There is more to Theorem 3 than one might suspect at first glance. In fact, Theorem 3 is equivalent to the compactness of a closed bounded interval (see Exercise 4). For suppose that $\mathcal{U}$ is an open cover of $I = [a, b]$. For each t there exists $U_t \in \mathcal{U}$ such that $t \in U_t$. Pick an open interval $\gamma(t)$ such that $t \in \gamma(t) \subset U_t$. Then γ is a gauge on I so there exists a γ-fine tagged partition $\mathcal{D} = \{(t_i, I_i) : 1 \leq i \leq n\}$. Then $I = \bigcup_{i=1}^n I_i \subset \bigcup_{i=1}^n \gamma(t_i) \subset \bigcup_{i=1}^n U_{t_i}$ so $\{U_{t_i} : 1 \leq i \leq n\}$ is a finite subcover of $\mathcal{U}$. For further applications, see in particular [M] and [Go5] where Theorem 3 is used to derive many of the fundamental results of real analysis such as the Intermediate Value Theorem, the Extreme Value Theorem and the uniform continuity theorem. See also [Bo1], [Bo2], [Sh] and [Th] for a result similar to Theorem 3 and applications.

In the remarks preceding the definition of the gauge integral, we indicated that the gauge integral can be viewed as being obtained from the Riemann integral by replacing the positive constant δ in the definition of the Riemann integral by a positive function δ. Indeed, it would be quite simple to restate Definition 2 by replacing the gauge γ by a positive function δ and employing condition (4) instead of (5). It is natural to ask how general the function δ can be in the resulting definition. This question has been addressed by Foran and Meinershagen ([FM]), Pfeffer ([Pf2]), Liu ([L]), and Gordon ([Go1] 9.24, [Go4]).

Finally, we give an example generalizing the result in Example 7 which will be used in several starred-sections.

Definition 9*. *A subset $E \subset [a,b]$ is (Lebesgue) null if for every $\varepsilon > 0$ there is a countable collection of open intervals $\{I_j\}$ covering E with $\sum_j \ell(I_j) < \varepsilon$.*

See Exercise 9 for several properties of null sets.

A property P concerning the points in a subset A is said to hold *almost everywhere* (abbreviated *a.e.*) in A if the property P holds for all points in A except those in a null set.

Example 10*. *Suppose $f : [a,b] \to \mathbb{R}$ is equal to 0 a.e. in I. Then f is integrable with $\int_I f = 0$. Let $\varepsilon > 0$ and $E = \{t : f(t) \neq 0\}$. For each $i \in \mathbb{N}$ let $E_i = \{t \in E : i - 1 < |f(t)| \leq i\}$. Then $E = \bigcup_{i=1}^{\infty} E_i$ and each E_i is null (Exercise 9a). For each i let $\{I_j^i : j \in \mathbb{N}\}$ be open intervals covering E_i with $\sum_{j=1}^{\infty} \ell(I_j^i) < \varepsilon/i2^i$. Define a gauge γ on I by $\gamma(t) = \mathbb{R}$ if $t \notin E$ and $\gamma(t) = I_j^i$ if $t \in E_i$ and j is the smallest integer such that $t_i \in I_j^i$. Suppose that $\mathcal{D} = \{(t_k, I_k) : 1 \leq k \leq m\} << \gamma$ and set $\mathcal{D}_i = \{(t_k, I_k) : t_k \in E_i\}$ for $i \in \mathbb{N}$ and $\mathcal{D}_0 = \{(t_k, I_k) : t \notin E\}$. Then $S(f, \mathcal{D}_0) = 0$ and*

$$|S(f, \mathcal{D}_i)| \leq \sum_{j=1}^{\infty} i\ell(I_j^i) < \varepsilon/2^i$$

for $i \geq 1$ so

$$|S(f, \mathcal{D})| \leq \sum_{i=1}^{\infty} \varepsilon/2^i = \varepsilon$$

and the result follows.

Exercises

1. Suppose γ_1 and γ_2 are gauges on I. Set $\gamma(t) = \gamma_1(t) \cap \gamma_2(t)$. Show that γ is a gauge on I such that any γ-fine tagged partition is also γ_1 and γ_2-fine.
2. Suppose γ_1 and γ_2 are gauges on I such that $\gamma_1(t) \subset \gamma_2(t)$ for all t. Show that any γ_1-fine tagged partition is also γ_2-fine.
3. Let $a < c < b$, $I = [a, b]$, $I_1 = [a, c]$, $I_2 = [c, b]$, and let γ be a gauge on I. If $\mathcal{D}_i$ is a γ-fine tagged partition of I_i for $i = 1, 2$, show $\mathcal{D} = \mathcal{D}_1 \cup \mathcal{D}_2 << \gamma$.
4. Use the following outline to give another proof of Theorem 3.
   Suppose the theorem is false and bisect the interval I. Use Exercise 3 to construct intervals $I_0 = I \supset I_1 \supset I_2 \supset \dots$ such that $\ell(I_k) \le \ell(I_{k-1})/2$ and no γ-fine tagged partition of I_k exists. Let $\{x\} = \bigcap_{k=1}^{\infty} I_k$ and obtain a contradiction.
5. Evaluate $\int_1^2 e^{-1/t}/t^2 dt$.
6. Suppose $g : I \to \mathbb{R}$ is non-negative and integrable and $f : I \to \mathbb{R}$ satisfies $|f(t)| \le g(t) \forall t \in I$. If $\int_I g = 0$, show that f is integrable over I and $\int_I f = 0$.
7. Suppose $f : I \to \mathbb{R}$ is such that $|f|$ is integrable over I with $\int_I |f| = 0$. Show that f is integrable over I with $\int_I f = 0$.
8. Let $a \le x_0 \le b$. Show that there is a gauge γ on $[a, b]$ such that if $\mathcal{D} << \gamma$ and J is the subinterval in $\mathcal{D}$ containing x_0, then x_0 must be the tag for J. Generalize this result to a finite number of points.

9*. Call a subset $E \subset [a, b]$ (Lebesgue) null if for every $\varepsilon > 0$ there is a countable collection of open intervals $\{I_j\}$ covering E with $\sum \ell(I_j) < \varepsilon$. (Definition 9)

   (a) Show that if E is null and $F \subset E$, then F is null.
   (b) Show that if each $\{E_j : j \in \mathbb{N}\}$ is null, then $\bigcup_{j \in \mathbb{N}} E_j$ is null.
   (c) Show that any countable set is null.
   (d) Show that if $E \subset [a, b]$ is null, then C_E is integrable with $\int_a^b C_E = 0$ where $C_E(t) = 1$ if $t \in E$ and $C_E(t) = 0$ otherwise.

10. (FTC). Let $f : [a, b] \to \mathbb{R}$ be continuous and have a derivative f' except for countably many points E in $[a, b]$. Define $\dot{f} : [a, b] \to \mathbb{R}$ by setting $\dot{f}(t) = f'(t)$ when $f'(t)$ exists and $\dot{f}(t) = 0$ otherwise. Show that $\dot{f}$ is integrable and $\int_a^b \dot{f} = f(b) - f(a)$. Show that the continuity assumption on f cannot be dropped. Given an example where Exercise 10 applies but Theorem 5 does not.
11. Show that in Definition 2 it can always be assumed that the tags are endpoints of the intervals which they tag. Show that if $\mathcal{D} << \gamma$, then there is a γ-fine tagged partition $\mathcal{D}'$ whose tags are endpoints and $S(f, \mathcal{D}) = S(f, \mathcal{D}')$.

Chapter 2

Basic Properties of the Gauge Integral

In this section we develop the basic properties of the (gauge) integral. Throughout this section let $I = [a, b]$ and $f, f_1, f_2, : I \to \mathbb{R}$.

Theorem 1. *Assume f_1 and f_2 are integrable over I.*

(i) *$f_1 + f_2$ is integrable over I with $\int_I (f_1 + f_2) = \int_I f_1 + \int_I f_2$.*
(ii) *For every $t \in \mathbb{R}$ tf_1 is integrable over I with $\int_I tf_1 = t\int_I f_1$.*
(iii) *If $f_1 \geq 0$ on I, then $\int_I f_1 \geq 0$.*
(iv) *If $f_1 \geq f_2$ on I, then $\int_I f_1 \geq \int_I f_2$.*

Proof:

(i): Let $\varepsilon > 0$. For $i = 1, 2$ there exist gauges γ_i such that $|S(f_i, \mathcal{D}) - \int_I f_i| < \varepsilon/2$ whenever $\mathcal{D} << \gamma_i$. Put $\gamma(t) = \gamma_1(t) \cap \gamma_2(t)$.

Then γ is a gauge and if $\mathcal{D} << \gamma$, then $\mathcal{D} << \gamma_i$ (Exercise 1.2) so

$$\left| S(f_1 + f_2, \mathcal{D}) - \left(\int_I f_1 + \int_I f_2 \right) \right|$$

$$\leq \left| S(f_1, \mathcal{D}) - \int_I f_1 \right| + \left| S(f_2, \mathcal{D}) - \int_I f_2 \right| < \varepsilon$$

and the result follows.

(ii): is left to Exercise 1.

(iii): Let $\varepsilon > 0$. There is a gauge γ_1 such that $|S(f_1, \mathcal{D}) - \int_I f_1| < \varepsilon$ whenever $\mathcal{D} << \gamma_1$. Since $f_1 \geq 0$,

$$0 \leq S(f_1, \mathcal{D}) \leq \int_I f_1 + \varepsilon$$

so $\int_I f_1 \geq 0$.

(iv): $f_1 - f_2 \geq 0$ on I so (iv) follows from (i), (ii) and (iii).

We say that f is *absolutely integrable* over I if both f and $|f|$ are integrable over I.

Corollary 2. *If f is absolutely integrable over I, then $|\int_I f| \leq \int_I |f|$.*

Proof: Since $f \leq |f|$ and $-f \leq |f|$ on I, Theorem 1 implies $\int_I f \leq \int_I |f|$ and $-\int_I f = \int_I(-f) \leq \int_I |f|$ so $|\int_I f| \leq \int_I |f|$.

In contrast to either the Riemann or Lebesgue integrals, we will see later (Example 12) that the absolute integrability assumption is important (see also Exercise 3.9).

An application of Theorem 1 and the FTC yields the familiar *integration by parts formula.*

Corollary 3. *Let f_1 and f_2 be differentiable over I. Then $f_1' f_2$ is integrable over I if and only if $f_1 f_2'$ is integrable over I and in this case*

$$\int_a^b f_1' f_2 = f_1(b) f_2(b) - f_1(a) f_2(a) - \int_a^b f_1 f_2' .$$

Proof: By the product rule we have $(f_1 f_2)' = f_1' f_2 + f_1 f_2'$. The result now follows from Theorems 1 and 1.5.

In the results above we have studied the basic properties of the integral $\int_I f$ as a function of the integrand f. We now consider the integral as a function of the interval I, i.e., we study the behavior of the integral as a set function.

Theorem 4. *Let $a < c < b$. If f is integrable over $[a, c]$ and $[c, b]$, then f is integrable over $[a, b]$ and*

$$\int_a^b f = \int_a^c f + \int_c^b f .$$

Proof: Let $\varepsilon > 0$. There exists a gauge $\gamma_1(\gamma_2)$ on $[a, c]([c, b])$ such that

$$|S(f, \mathcal{D}_1) - \int_a^c f| < \varepsilon/2$$

whenever $\mathcal{D}_1$ is a γ_1-fine tagged partition of $[a,c]$($|S(f,\mathcal{D}_2) - \int_c^b f| < \varepsilon/2$ whenever $\mathcal{D}_2$ is a γ_2-fine tagged partition of $[c,b]$). Define a gauge γ on $[a,b]$ by setting

$$\gamma(t) = \begin{cases} (a,c) \cap \gamma_1(t) \text{ if } t \in (a,c)\,, \\ (c,b) \cap \gamma_2(t) \text{ if } t \in (c,b)\,, \\ \gamma_1(c) \cap \gamma_2(c) \text{ if } t = c\,, \\ \gamma_1(a) \cap (-\infty, c) \text{ if } t = a\,, \\ \gamma_2(b) \cap (c, \infty) \text{ if } t = b\,. \end{cases}$$

If $\mathcal{D}$ is a γ-fine tagged partition of $[a,b]$, then $\mathcal{D}$ contains either one subinterval J with c as a tag or $\mathcal{D}$ contains two subintervals with c as a tag. (Note that c must be a tag.) In the former case, we can "divide" J at c without changing the Riemann sum $S(f,\mathcal{D})$ and, therefore, obtain the latter case. In the latter case $\mathcal{D}_1 = \{(t,J) \in \mathcal{D} : J \subset [a,c]\}$ and $\mathcal{D}_2 = \{(t,J) \in \mathcal{D} : J \subset [c,b]\}$ are tagged partitions of $[a,c]$ and $[c,b]$, respectively, with $\mathcal{D}_i << \gamma_i$. Then

$$\left|S(f,\mathcal{D}) - \left(\int_a^c f + \int_c^b f\right)\right| \le \left|S(f,\mathcal{D}_1) - \int_a^c f\right| + \left|S(f,\mathcal{D}_2) - \int_c^b f\right| < \varepsilon$$

so the result follows.

Remark 5. *If J is a subinterval of $[a,b]$ and f is the characteristic function of J it follows from Example* 1.7 *and Theorem* 4 *that f is integrable over $[a,b]$ with $\int_I f = \ell(J)$. Thus, if f is a step function $f = \sum_{j=1}^n a_j C_{A_j}$, where $a_j \in \mathbb{R}$, A_j is a subinterval of I and C_{A_j} is the characteristic function of A_j, then f is integrable ovcr I with*

$$\int_I f = \sum_{j=1}^n a_j \ell(A_j)\,.$$

To establish the converse of Theorem 4 we establish the Cauchy criterion for the integral. Similar to the *Cauchy criterion* for sequences of real numbers, this criterion relieves us of the responsibility of having a value for the integral in hand.

Theorem 6. *Let $f : I \to \mathbb{R}$. Then f is integrable over I if and only if for every $\varepsilon > 0$ there is a gauge γ on I such that if $\mathcal{D}_1, \mathcal{D}_2 << \gamma$, then $|S(f,\mathcal{D}_1) - S(f,\mathcal{D}_2)| < \varepsilon$.*

Proof: That the condition is necessary follows from the triangle inequality.

For sufficiency note that for every k there is a gauge γ_k on I such that if $\mathcal{D}_1, \mathcal{D}_2 << \gamma_k$, then

$$|S(f, \mathcal{D}_1) - S(f, \mathcal{D}_2)| < 1/k\,.$$

We may assume that $\gamma_1(t) \supset \gamma_2(t) \supset \ldots$ for every $t \in I$ (Exercise 1.1). For each k let $\mathcal{D}_k << \gamma_k$. If $k > j$, we have $|S(f, \mathcal{D}_k) - S(f, \mathcal{D}_j)| < 1/j$. Thus, $\{S(f, \mathcal{D}_k)\}$ is a Cauchy sequence in $\mathbb{R}$; let $A = \lim S(f, \mathcal{D}_k)$. Then

$$|S(f, \mathcal{D}_k) - A| \le 1/k$$

for every k.

Let $\varepsilon > 0$ and pick N such that $1/N < \varepsilon/2$. Suppose $\mathcal{D} << \gamma_N$. Then

$$\begin{aligned}|S(f, \mathcal{D}) - A| &\le |S(f, \mathcal{D}) - S(f, \mathcal{D}_N)| + |S(f, \mathcal{D}_N) - A| \\ &< 1/N + 1/N < \varepsilon\,.\end{aligned}$$

Thus, f is integrable over I with $\int_I f = A$.

We can now establish the converse of Theorem 4.

Theorem 7. *Let $f : I \to \mathbb{R}$ be integrable over I. If J is a closed subinterval of I, then f is integrable over J.*

Proof: Since we have no candidate for $\int_J f$, we use the Cauchy Criterion of Theorem 6. Let $\varepsilon > 0$. There is a gauge γ on I such that if $\mathcal{D}_1, \mathcal{D}_2$ are γ-fine tagged partitions of I, then $|S(f, \mathcal{D}_1) - S(f, \mathcal{D}_2)| < \varepsilon$.

Consider the case $a < c < d < b$ and $J = [c, d]$; the other cases are similar. Let γ' be the restriction of γ to J and let $\gamma_1(\gamma_2)$ be the restriction of γ to $[a, c]([c, b])$. Let $\mathcal{D}_1(\mathcal{D}_2)$ be a $\gamma_1(\gamma_2)$-fine tagged partition of $[a, c]([c, b])$. Now suppose $\mathcal{D}$ and $\mathcal{E}$ are γ'-fine tagged partitions of J. Then $\mathcal{D}' = \mathcal{D}_1 \cup \mathcal{D} \cup \mathcal{D}_2 << \gamma$ and $\mathcal{E}' = \mathcal{D}_1 \cup \mathcal{E} \cup \mathcal{D}_2 << \gamma$ so

$$|S(f, \mathcal{D}') - S(f, \mathcal{E}')| = |S(f, \mathcal{D}) - S(f, \mathcal{E})| < \varepsilon\,.$$

It follows from Theorem 6 that f is integrable over J.

From Theorems 4 and 7 a function $f : I \to \mathbb{R}$ is integrable over I if and only if f is integrable over every closed subinterval of I. Moreover, if $\{I_i : 1 \le i \le n\}$ is a partition of I and f is integrable over I, $\int_I f = \sum_{i=1}^n \int_{I_i} f$; this property of the integral is usually referred to as (finite) *additivity* for the integral.

In Chapter 1 we proved what we referred to as the Fundamental Theorem of Calculus (FTC). Actually, we proved only one half of what is usually referred to as the FTC, the half concerning the integration of a derivative. The other half concerns the differentiation of the indefinite integral. Using Theorems 1 and 7 we can now address this half of the FTC.

Theorem 8 (*FTC: Part 2*). *Let $f : I \to \mathbb{R}$ be integrable over I and set $F(x) = \int_a^x f$ for $a \le x \le b$. If f is continuous at $x \in [a, b]$, then F is differentiable at x with $F'(x) = f(x)$.*

Proof: Let $\varepsilon > 0$. There exists $\delta > 0$ such that $|f(x) - f(t)| < \varepsilon$ for $|x - t| < \delta, t \in [a, b]$. Suppose $|y - x| < \delta, y \in [a, b]$ and $y > x$. Then

$$f(x) - \varepsilon < f(t) < f(x) + \varepsilon$$

for $x \le t \le y$. Using Theorems 1 and 7, integration over $[x, y]$ yields

$$(f(x) - \varepsilon)(y - x) \le \int_x^y f(t)dt \le (f(x) + \varepsilon)(y - x)$$

or

$$f(x) - \varepsilon \le (F(y) - F(x))/(y - x) \le f(x) + \varepsilon\,.$$

Since a similar inequality holds for $y < x$, the result follows.

See the Notes/Remarks section for references to more general results.

We can employ the Cauchy Criterion of Theorem 6 to show that any continuous function is integrable over a closed, bounded interval. For this we first establish a lemma which is interesting in its own right.

Lemma 9. *Let $f : I \to \mathbb{R}$ and assume that for every $\varepsilon > 0$ there exist integrable functions $g_1, g_2 : I \to \mathbb{R}$ such that $g_1 \le f \le g_2$ on I and $\int_I g_2 \le \int_I g_1 + \varepsilon$. Then f is integrable over I.*

Proof: Let $\varepsilon > 0$. By Exercise 1.1 there is a gauge γ on I such that if $\mathcal{D} << \gamma$, then $|S(g_i, \mathcal{D}) - \int_I g_i| < \varepsilon$ for $i = 1, 2$. Suppose $\mathcal{D} << \gamma$. Then

$$\int_I g_1 - \varepsilon < S(g_1, \mathcal{D}) \le S(f, \mathcal{D}) \le S(g_2, \mathcal{D}) < \int_I g_2 + \varepsilon \le \int_I g_1 + 2\varepsilon\,.$$

Thus any Riemann sum for f with respect to a γ-fine tagged partition lies within an interval with endpoints $[\int_I g_1 - \varepsilon, \int_I g_1 + 2\varepsilon]$ so any two such Riemann sums differ by at most 3ε. It follows from Theorem 6 that f is integrable over I.

Theorem 10. *If $f : I \to \mathbb{R}$ is continuous, then f is integrable over I.*

Proof: Let $\varepsilon > 0$. Since f is uniformly continuous on I, there is a $\delta > 0$ such that $|f(x) - f(y)| < \varepsilon$ when $x, y \in I, |x - y| < \delta$. Let $\mathcal{P} = \{a = x_0 < \cdots < x_n = b\}$ be a partition of $[a, b]$ such that $(x_i - x_{i-1}) < \delta$ for $i = 1, \ldots, n$. For $i = 1, \ldots, n$ put $M_i = \sup\{f(t) : x_{i-1} \le t \le x_i\}$ and $m_i = \inf\{f(t) : x_{i-1} \le t \le x_i\}$ and define step functions g_1 and g_2 by

$$g_1 = m_1 C_{[x_0,x_1]} + \sum_{i=2}^{n} m_i C_{(x_{i-1},x_i]},\ g_2 = M_1 C_{[x_0,x_1]} + \sum_{i=2}^{n} M_i C_{(x_{i-1},x_i]}\,.$$

Then $g_1 \le f \le g_2$ and $0 \le g_2 - g_1 \le \varepsilon$ on I so $\int_I (g_2 - g_1) \le \varepsilon(b - a)$. Lemma 9 now gives the result.

We can employ Theorem 10 to obtain a version of the Mean Value Theorem for integrals.

Corollary 11. *Let $f : I \to \mathbb{R}$ be continuous. There exists $t \in I$ such that*

$$\int_a^b f = f(t)(b - a)\,.$$

Proof: Let $M = \max\{f(t) : t \in I\}, m = \inf\{f(t) : t \in I\}$. Then

$$m(b - a) \le \int_I f \le M(b - a)$$

or

$$m \le \int_I f/(b - a) \le M\,.$$

The result now follows from the Intermediate Value Theorem for continuous functions.

We can use the basic properties of the integral developed above along with the FTC to give an example illustrating a property of the gauge integral which it does not share with the Riemann or Lebesgue integrals. Namely, the gauge integral is a "conditional" integral; that is, it admits functions which are integrable but whose absolute values are not integrable.

Example 12. *For $0 < t \le 1$, let $f(t) = t^2 \cos(\pi/t^2)$ and $f(0) = 0$. Then f is differentiable on $[0, 1]$ with*

$$f'(t) = \begin{cases} 0 & t = 0 \\ 2t\cos(\pi/t^2) + \dfrac{2\pi}{t}\sin(\pi/t^2) & 0 < t \le 1\,. \end{cases}$$

By Theorem 1.5 f' *is integrable over* [0,1] *with* $\int_0^1 f' = -1$. *However,* $|f'|$ *is not integrable over* $[0,1]$. *To see this let* $\alpha_k = \sqrt{2/(4k+1)}$ *and* $\beta_k = 1/\sqrt{2k}$. *Then the intervals* $\{[\alpha_k, \beta_k]\}$ *are pairwise disjoint and* $|f'|$ *is integrable over each* $[\alpha_k, \beta_k]$ *by Theorem* 10. *By the FTC and Corollary* 2, *we have*

$$\int_{\alpha_k}^{\beta_k} |f'| \geq \left| \int_{\alpha_k}^{\beta_k} f' \right| = 1/2k.$$

If $|f'|$ *is integrable over* $[0,1]$, *we have from the finite additivity of the integral*

$$\int_0^1 |f'| \geq \sum_{k=1}^n \int_{\alpha_k}^{\beta_k} |f'| \geq \sum_{k=1}^n 1/2k$$

for every n *which is clearly impossible.*

If $f : I \to \mathbb{R}$ is integrable over I and $|f|$ is also integrable over I, we say that f is *absolutely integrable* over I. If f is integrable over I but $|f|$ fails to be integrable over I, we say that f is *conditionally integrable* over I. Thus, the function in Example 12 is an example of a function which is conditionally integrable.

Notes/Remarks

There is another version of part 2 of the FTC which implies that the indefinite integral of an integrable function is differentiable at "most" points in $[a, b]$ with derivative equal to f. This result is established in Appendix 3. This result also establishes another important property of an integrable function; namely, that any gauge integrable function is (Lebesgue) measurable. For other proofs of the measurability, see [L], [LPY], [LPY1].

There are also known characterizations of the indefinite gauge integral; e.g., [Gol] 9.17, [LPY] 6.22.

The computations in Example 12 show that the FTC does not hold for the Lebesgue integral since the Lebesgue integral is an absolute integral.

There are several characterizations of gauge integrability in terms of the Lebesgue integral. First, Liu has given the following necessary condition for gauge integrability. ([Liu]; see also [LCL].)

Theorem 13. *If* f *is gauge integrable on* $[a, b]$, *then there is an increasing sequence of closed subsets* $\{X_k\}$ *of* $[a, b]$ *such that* $\bigcup_{k=1}^{\infty} X_k = [a, b]$, f *is Lebesgue integrable on each* X_k *and* $\lim_k L\int_{X_k} f = \int_a^b f$, *where* $L\int$ *denotes the Lebesgue integral.*

A necessary and sufficient condition for gauge integrability is given by

Theorem 14 (*LCL*). *$f : [a, b] \to \mathbb{R}$ is gauge integrable over $[a, b]$ if and only if there is an increasing sequence of closed subsets $\{X_k\}$ of $[a, b]$ such that $\bigcup_{k=1}^{\infty} X_k = [a, b]$, f is Lebesgue integrable over each X_k and the following condition holds:*
for every $\varepsilon > 0$ there exists N such that if $k \geq N$, then there exists a gauge γ_k on $[a, b]$ such that whenever $\mathcal{D} = \{(t_i, I_i) : 1 \leq i \leq m\} << \gamma_k$, then $|\sum_{t_i \notin X_k} f(t_i)\ell(I_i \cap J)| < \varepsilon$ for every subinterval J.

Other characterizations of gauge integrability can be found in [Sc] and [LA1].

Another interesting connection between the gauge and Lebesgue integrals is given by

Theorem 15. *Let $f : [a, b] \to \mathbb{R}$ be gauge integrable over $[a, b]$. If f is gauge integrable over every measurable subset of $[a, b]$ (i.e., if $C_E f$ is gauge integrable over $[a, b]$ for every measurable $E \subset [a, b]$), then f is Lebesgue integrable over $[a, b]$.*

See [Go1] p. 146 for a proof.

Exercises

1. Prove Theorem 1 (ii).
2. Evaluate $\int_0^1 xe^{-x}dx$.
3. Let $f : I \to \mathbb{R}$. Suppose $\exists A \in \mathbb{R}$ such that for every $\varepsilon > 0$ there are integrable functions g and h with $g \leq f \leq h$ and $A - \varepsilon < \int_I g \leq \int_I h < A + \varepsilon$. Show that f is integrable with $\int_I f = A$.
4. Let $f : I \to \mathbb{R}$ be integrable over I and suppose that $g : I \to \mathbb{R}$ is equal to f except possibly at countably many points in I. Show that g is integrable with $\int_I g = \int_I f$.
5. Let $f(t) = \sin t$ if $t \in [0, 1] \backslash \mathbb{Q}$ and $f(t) = t$ if $t \in \mathbb{Q} \cap [0, 1]$. Show that f is integrable over $[0, 1]$ and calculate $\int_0^1 f$.
6. Can the condition $f_1(t) \geq f_2(t)$ for all $t \in I$ in Theorem 1 (iv) be relaxed?
7. Suppose $\int_I |f| = 0$. Show that f is integrable over I with $\int_I f = 0$.
8. Suppose $\int_I |f - g| = 0$. Show that f is integrable over I if and only if g is integrable over I with $\int_I f = \int_I g$.
9. Suppose $f : I \to \mathbb{R}$ is non-negative and continuous on I. If $\int_I f = 0$, show that $f \equiv 0$ on I. Can the non-negativity assumption be dropped? How about the continuity?

10. Let $f, g : I \to \mathbb{R}$ be continuous with $g \geq 0$. Show that $\exists\ t \in I$ such that $\int_a^b fg = f(t) \int_a^b g$.
11. Let $f(t) = 1$ for $0 \leq t \leq 1$ and $f(t) = -1$ for $-1 \leq t < 0$. Is the indefinite integral F of f differentiable at $t = 0$? Why doesn't this violate Theorem 8?
12. (Linear Change of Variable) Let $f : [a, b] \to \mathbb{R}$ be integrable over $[a, b]$ and let $h \in \mathbb{R}$. Define $f_h : [a + h, b + h] \to \mathbb{R}$ by $f_h(t) = f(t - h)$. Show that f_h is integrable with $\int_a^b f = \int_{a+h}^{b+h} f_h$.
13. Let $f : \mathbb{R} \to \mathbb{R}$ be continuous and set $f_k(x) = \int_x^{x+1/k} f$. Show that $f_k \to f$ pointwise on $\mathbb{R}$ and the convergence is uniform if f is uniformly continuous.
14. If $f : [a, b] \to \mathbb{R}$ is continuous, $|f| \leq 1$ and $\int_a^b f = (b - a)$, what is f?
15. Let $f : [a, b] \to \mathbb{R}$ be continuously differentiable with $f(a) = f(b) = 0$. Show that if $\int_a^b f^2 = 1$, then $\int_a^b tf(t)f'(t)dt = -1/2$.
16. Let $f : [a, b] \to \mathbb{R}$ be continuous on $[a, b]$ and differentiable on (a, b). If $f(a) = 0$, $f(b) = -1$ and $\int_a^b f = 0$, show that $\exists\ c \in (a, b)$ such that $f'(c) = 0$.
17. Let $f : [0, 1] \to \mathbb{R}$ be continuous. Let $f_0 = f$ and $f_{k+1}(t) = \int_0^t f_k$ for $k \geq 0$. Show that if $f_k = 0$ for some k, then $f = 0$.
18. (Dilation) Let $f : [a, b] \to \mathbb{R}$ be integrable over $[a, b]$. For $c \neq 0$ let $g(t) = f(ct)$, $a/c \leq t \leq b/c$. Show that g is integrable over $[a/c, b/c]$ with $\int_a^b f(t)dt = c \int_{a/c}^{b/c} g(t)dt$.
19. Show that continuity cannot be dropped in Corollary 11.
20. Let $f : [0, 1] \to \mathbb{R}$ be continuous and increasing. If $g(x) = (1/x) \int_0^x f$ for $0 < x \leq 1$, show that g is increasing.
21. Let $f : [a, b] \to \mathbb{R}$ be positive and continuous with $M = \max\{f(t) : a \leq t \leq b\}$. Show that $M = \lim(\int_a^b f(t)^n dt)^{1/n}$. [Hint: $f \geq M - \varepsilon$ on some subinterval.]

22*. Let $f : I \to \mathbb{R}$ be integrable over $\mathbb{R}$ and equal to $g : I \to \mathbb{R}$ a.e. Show that g is integrable over I with $\int_I f = \int_I g$. [See Definition 1.9.]

Chapter 3

Henstock's Lemma and Improper Integrals

One of the most important properties of the gauge integral is the validity of convergence theorems of the form, $\lim \int_I f_k = \int_I (\lim f_k)$, under rather general hypotheses. Such general convergence theorems hold for the Lebesgue integral and are the principal reason that the Lebesgue integral is so much superior to the Riemann integral. We will establish these convergence theorems in Chapter 5. The principal tool used in establishing the convergence theorems is a result usually referred to as *Henstock's Lemma.* We begin with this result and then give some applications. First we need some terminology.

Let $I = [a, b]$. A *partial tagged partition* of I is a finite collection of pairs

$$\mathcal{J} = \{(t_i, J_i) : 1 \leq i \leq n\},$$

where the $\{J_i\}$ are non-overlapping closed subintervals of I and $t_i \in J_i$ (it is not required that $\bigcup_{i=1}^n J_i = I$). If γ is a gauge on I and $\mathcal{J}$ is a partial tagged partition of I, then $\mathcal{J}$ is said to be *γ-fine*, written $\mathcal{J} << \gamma$, if $t_i \in J_i \subset \gamma(t_i)$ for $i = 1, \ldots, n$. If $\mathcal{J}$ is a partial tagged partition of I and $f : I \to \mathbb{R}$, we write $S(f, \mathcal{J}) = \sum_{i=1}^n f(t_i)\ell(J_i)$ for the *Riemann sum* of f with respect to $\mathcal{J}$, and if $J = \bigcup_{i=1}^n J_i$, we write $\int_J f = \sum_{i=1}^n \int_{J_i} f$ when f is integrable. In case $\mathcal{D}$ is a (γ-fine) tagged partition of I, this agrees with our previous definitions.

Lemma 1 (*Henstock*). *Let $f : I \to \mathbb{R}$ be integrable over I. For $\varepsilon > 0$ suppose γ is a gauge on I such that if $\mathcal{D}$ is a γ-fine tagged partition of I, then $|S(f, \mathcal{D}) - \int_I f| < \varepsilon$. If $\mathcal{J} = \{(t_i, J_i) : 1 \leq i \leq n\}$ is any partial tagged partition of I such that $\mathcal{J} << \gamma$, then*

$$\left| S(f, \mathcal{J}) - \int_J f \right| \leq \varepsilon,$$

where $J = \bigcup_{i=1}^{n} J_i$, *and*

$$\sum_{i=1}^{n} \left| f(t_i)\ell(J_i) - \int_{J_i} f \right| \le 2\varepsilon .$$

Proof: The set $I \setminus \bigcup_{i=1}^{n} J_i$ consists of a finite number of disjoint intervals. Let K_i, $1 \le i \le m$, be the closures of these subintervals. For $\eta > 0$ the integrability of f over each K_i implies that there exists a γ-fine tagged partition $\mathcal{K}_i$ of K_i such that

$$\left| S(f, \mathcal{K}_i) - \int_{K_i} f \right| < \eta/m .$$

Then $\mathcal{D} = \mathcal{J} \cup \mathcal{K}_1 \cup \ldots \cup \mathcal{K}_m$ is a γ-fine tagged partition of I. If $J = \bigcup_{i=1}^{n} J_i$, then

$$\left| S(f, \mathcal{D}) - \int_I f \right| = \left| S(f, \mathcal{J}) - \int_J f + \sum_{i=1}^{m} \left(S(f, \mathcal{K}_i) - \int_{K_i} f \right) \right| < \varepsilon$$

so

$$\left| S(f, \mathcal{J}) - \int_J f \right| < \varepsilon + \sum_{i=1}^{m} \left| S(f, \mathcal{K}_i) - \int_{K_i} f \right| < \varepsilon + m(\eta/m) = \varepsilon + \eta .$$

Since $\eta > 0$ is arbitrary, $|S(f, \mathcal{J}) - \int_J f| \le \varepsilon$ as desired.

For the second inequality, let $\mathcal{J}_+ (\mathcal{J}_-)$ be those (t_i, J_i) such that $f(t_i)\ell(J_i) - \int_{J_i} f \ge 0$ (< 0). Then by the first inequality

$$0 \le \sum_{\mathcal{J}_+} \left(f(t_i)\ell(J_i) - \int_{J_i} f \right) = \sum_{\mathcal{J}_+} \left| f(t_i)\ell(J_i) - \int_{J_i} f \right| \le \varepsilon$$

and

$$0 \le -\sum_{\mathcal{J}_-} \left(f(t_i)\ell(I_i) - \int_{J_i} f \right) = \sum_{\mathcal{J}_-} \left| f(t_i)\ell(J_i) - \int_{J_i} f \right| \le \varepsilon$$

so the second inequality follows.

Henstock's Lemma asserts that if γ is a gauge on I such that γ-fine tagged partitions of I induce Riemann sums which give good approximations to the value of the integral over I, then likewise any γ-fine partial tagged partition induces Riemann sums which give good approximations to the value of the integral over the union of the intervals in the partial tagged partition.

We now give some immediate applications of Henstock's Lemma.

Corollary 2. *Let $f : I \to \mathbb{R}$ be integrable over I and set $F(x) = \int_a^x f$ for $a \le x \le b$. Then F is continuous on I.*

Proof: Let $\varepsilon > 0$ and $x \in I$. There is a gauge γ on I such that $|S(f, \mathcal{D}) - \int_I f| < \varepsilon$ whenever $\mathcal{D} << \gamma$. Suppose $\gamma(x) = (\alpha, \beta)$ and set $\delta = \min\{\beta - x, x - \alpha, \varepsilon/(1 + |f(x)|)\}$. Now suppose that $|y - x| < \delta$ with $y \in I$. Then from Henstock's Lemma applied to $\{(x, [y, x])\}$ or $\{(x, [x, y])\}$ we have

$$\left| f(x)(y - x) - \int_x^y f \right| \le \varepsilon \text{ or } \left| f(x)(x - y) - \int_y^x f \right| \le \varepsilon .$$

In either case $|\int_x^y f| = |F(y) - F(x)| \le \varepsilon + |f(x)||y - x| \le \varepsilon + \varepsilon = 2\varepsilon$ so F is continuous at x.

We give an application of the second inequality in Henstock's Lemma.

Corollary 3. *Let $f : I \to \mathbb{R}$ be integrable over I. If $\int_a^c f = 0$ for every $c \in [a, b]$, then $|f|$ is integrable on I with $\int_I |f| = 0$.*

Proof: Note that if $a < c < d \le b$, then $\int_c^d f = \int_a^d f - \int_a^c f = 0$. Let $\varepsilon > 0$ and let γ be a gauge on I such that

$$\left| S(f, \mathcal{D}) - \int_I f \right| = |S(f, \mathcal{D})| < \varepsilon$$

whenever

$$\mathcal{D} = \{(t_i, I_i) : 1 \le i \le n\} << \gamma .$$

By Henstock's Lemma

$$\sum_{i=1}^{n} |f(t_i)| \ell(I_i) \le 2\varepsilon .$$

This implies that $|f|$ is integrable with $\int_I |f| = 0$.

Finally, we conclude this section by considering improper integral for the gauge integral.

Theorem 4. *Let $f : [a, b] \to \mathbb{R}$ be integrable over $[c, b]$ for every $a < c < b$. Then f is integrable over $[a, b]$ if and only if $\lim_{c \to a+} \int_c^b f = A$ exists. In this case $\int_a^b f = A$.*

Proof: For the necessity, let $\varepsilon > 0$ and choose a gauge γ on $[a,b]$ such that

$$\left|S(f,\mathcal{D}) - \int_a^b f\right| < \varepsilon$$

whenever $\mathcal{D} << \gamma$. For each $c \in (a,b)$ there is a gauge γ_c on $[c,b]$ such that

$$\left|S(f,\mathcal{E}) - \int_c^b f\right| < \varepsilon$$

whenever $\mathcal{E}$ is a γ-fine tagged partition of $[c,b]$. We may assume $\gamma_c(t) \subset \gamma(t)$ for all $t \in [c,b]$. Choose c such that $c \in \gamma(a)$ and $(c-a)|f(a)| < \varepsilon$.

Suppose $s \in (a,c)$. Let $\mathcal{E}$ be a γ_s-fine tagged partition of $[s,b]$ and set $\mathcal{D} = \mathcal{E} \cup \{(a,[a,s])\}$. Note $\mathcal{D} << \gamma$ so

$$\left|\int_a^b f - \int_s^b f\right| \le \left|\int_a^b f - S(f,\mathcal{D})\right| + \left|S(f,\mathcal{E}) - \int_s^b f\right| + |f(a)||s-a| < 3\varepsilon\,.$$

Hence, $\lim_{c\to a+} \int_c^b f = \int_a^b f$.

For sufficiency, let $\{c_k : k = 0,1,\dots\} \subset (a,b]$ be such that $c_0 = b$, $c_k > c_{k+1}$ and $c_k \to a$. Pick a gauge γ_1 on $[c_1,c_0]$ such that $|S(f,\mathcal{D}) - \int_{c_1}^{c_0} f| < \varepsilon/2$ whenever $\mathcal{D}$ is a γ_1-fine tagged partition of $[c_1,c_0]$. For $k \ge 2$ pick a gauge γ_k on $[c_k,c_{k-2}]$ such that

$$\left|S(f,\mathcal{D}) - \int_{c_k}^{c_{k-2}} f\right| < \varepsilon/2^k$$

whenever $\mathcal{D}$ is a γ_k-fine tagged partition of $[c_k,c_{k-2}]$. Choose N such that $|A - \int_s^{c_0} f| < \varepsilon$ for $a < s \le c_N$ and $|f(a)|(c_N - a) < \varepsilon$. Now define a gauge γ by setting

$$\gamma(t) = \begin{cases} (-\infty, c_N) & t = a \\ \gamma_1(t) \cap (c_1,\infty) & c_1 < t \le c_0 \\ \gamma_k(t) \cap (c_k, c_{k-2}) & c_k < t \le c_{k-1}, k \ge 2\,. \end{cases}$$

Now suppose $\mathcal{D} = \{(t_i, I_i) : 1 \le i \le n\} << \gamma$. For each k let $\mathcal{D}_k$ be the subset of $\mathcal{D}$ whose tags are in $(c_k, c_{k-1}]$ (only a finite number of the $\mathcal{D}_k$ are non-empty and no 2 $\mathcal{D}_k$'s have common elements).

Let J_k be the union of the subintervals belonging to $\mathcal{D}_k$. By the definition of γ_k each $\mathcal{D}_k << \gamma_k$, $J_1 \subset (c_1, c_0]$ and $J_k \subset (c_k, c_{k-2})$ for $k \ge 2$. By

Henstock's Lemma $|S(f,\mathcal{D}_k) - \int_{J_k} f| \le \varepsilon/2^k$. Note that if $(x,K) \in \mathcal{D}$ is such that the subinterval $K = [a,d]$ is the subinterval in $\mathcal{D}$ containing a, then the tag associated with K must be a, i.e., $a = x$. [Assume that $a < x$. Then $a < c_k < x$ for some k and $x \in K \subset \gamma(x) \subset (c_k, c_{k-2})$. This is impossible since $a \in K$.] Thus,

$$|A - S(f,\mathcal{D})| \le |f(a)|\ell(K) + \left|\sum_{k=1}^{\infty}\left(\int_{J_k} f - S(f,\mathcal{D}_k)\right)\right|$$

$$+ \left|A - \int_d^b f\right| < \varepsilon + \sum_{k=1}^{\infty} \varepsilon/2^k + \varepsilon = 3\varepsilon\,.$$

It follows that there are no "improper" or "Cauchy-Riemann" integrals for the gauge integral over bounded intervals. We will show in Chapter 5 that a similar situation occurs for the gauge integral over unbounded intervals. This is in sharp contrast to the situation for both the Riemann and Lebesgue integrals.

We can employ Theorem 4 to obtain a "comparison" criterion for the existence of integrals.

Corollary 5. *Let $f, g : [a,b] \to \mathbb{R}$ with $|f| \le g$ on $[a,b]$. Suppose f is absolutely integrable over $[c,b]$ for $a < c \le b$.*

(i) *If f is non-negative, then f is integrable over $[a,b]$ if and only if $\{\int_c^b f : a < c \le b\}$ bounded.*
(ii) *If g is integrable over $[a,b]$, then f is integrable over $[a,b]$.*

Proof: (i): Note the function $c \to \int_c^b f$ is decreasing on $(a,b]$ so Theorem 4 gives the result immediately.

(ii): Put $F(s) = \int_s^b f, G(s) = \int_s^b g$ for $a < s \le b$. By Theorem 4 it suffices to show that the function F satisfies a Cauchy criterion near a. For $a < s < t$,

$$|F(t) - F(s)| \le \int_s^t |f| \le \int_s^t g = G(t) - G(s)\,.$$

But, by Theorem 4, the function G satisfies a Cauchy criterion near a so f is integrable over $[a,b]$.

Later, in Corollary 4.11 it will be shown that the absolute integrability assumption in Corollary 5 can be relaxed to integrability.

We conclude this section with two examples.

Example 6. *For $p \in \mathbb{R}$ let $f(t) = t^p, 0 \le t \le 1$. For $0 < c < 1$ and $p \ne -1$,*

$$\int_c^1 f = (1 - c^{p+1})/(p+1)\,.$$

This expression has a finite limit as $c \to 0$ if and only if $p+1 > 0$, and in this case we have $\int_0^1 t^p dt = \frac{1}{p+1}$ from Theorem 4. For $p = -1$, we have $\int_c^1 f = -\ln c$ so t^{-1} is not integrable over $[0,1]$ by Theorem 4. Thus, t^p is integrable over $[0,1]$ if and only if $p > -1$.

We next consider the beta function, often encountered in probability and statistics.

Example 7. *The beta function is defined to be*

$$B(x,y) = \int_0^1 t^{x-1}(1-t)^{y-1} dt\,.$$

For $x \ge 1$ and $y \ge 1$ the integrand is continuous so the integral exists. However, for $x < 1 (y < 1)$ the integrand has a singularity at $t = 0 (t = 1)$. By splitting the integral into two parts, $\int_0^{1/2} + \int_{1/2}^1$, we may consider these cases separately. Now $(1-t)^{y-1}$ is bounded on $[0,1/2]$ and t^{x-1} is integrable on $[0,1/2]$ for $x > 0$ by Example 6. From Corollary 5 $\int_0^{1/2} t^{x-1}(1-t)^{y-1}dt$ exists for $x > 0$. The change of variable $s = 1 - t$ can be used to show that similarly $\int_{1/2}^1 t^{x-1}(1-t)^{y-1}dt$ exists for $y > 0$. Thus, the beta function is defined for $x > 0, y > 0$.

Notes/Remarks

Henstock's Lemma (3.1) is sometimes referred to as the Saks-Henstock Lemma; Henstock indicates that he adapted Sak's proof of a result for the Burkill integral ([H3] p. 197; [Sa] p. 214). The lemma is used later in the proofs of the Monotone and Dominated Convergence Theorems.

Lee Tuo-Yeong, Chew Tuan-Seng and Lee Peng Yee have established a uniform version of the Henstock Lemma and employed the lemma to give a necessary and sufficient condition for a function to be gauge integrable in terms of the Lebesgue integral ([LCL] Lemma 3). We give their result. It is used later in Chapter 7.

Lemma 8 (*Uniform Henstock Lemma*). *Let $f : I \to \mathbb{R}$ be integrable over I and $\varepsilon > 0$. Suppose γ is a gauge on I such that $|S(f,\mathcal{D}) - \int_I f| < \varepsilon$ whenever*

$\mathcal{D} << \gamma$. *If J is any subinterval of I, then*

$$\left|\sum_{i=1}^{m}\left\{f(t_i)\ell(I_i\cap J)-\int_{I_i\cap J} f\right\}\right|\le 3\varepsilon$$

and

$$\sum_{i=1}^{m}\left|f(t_i)\ell(I_i\cap J)-\int_{I_i\cap J} f\right|\le 6\varepsilon$$

whenever $\mathcal{D}=\{(t_i,I_i):1\le i\le m\}$ is a γ-fine partial tagged partition of I.

Proof: Let $\mathcal{D}=\{(t_i,I_i):1\le i\le m\}$ be a γ-fine partial tagged partition of I and set

$$d_0=\{i:1\le i\le m,\ \ell(I_i\cap J)>0\},\ \mathcal{D}_0=\{(t_i,I_i\cap J):i\in d_0\}.$$

If

$$S=\left|\sum_{i=1}^{m}\left\{f(t_i)\ell(I_i\cap J)-\int_{I_i\cap J} f\right\}\right|,$$

then

$$S=\left|\sum_{i\in d_0}\left\{f(t_i)\ell(I_i\cap J)-\int_{I_i\cap J} f\right\}\right|,$$

and we want to show $S\le 3\varepsilon$. If

$$d_1=\{i\in d_0:t_i\in I_i\cap J\},$$

then $\mathcal{D}_1=\{(t_i,I_i\cap J):i\in d_1\}$ is γ-fine so

$$\left|\sum_{i\in d_1}\left\{f(t_i)\ell(I_i\cap J)-\int_{I_i\cap J} f\right\}\right|\le\varepsilon$$

by Henstock's Lemma (3.1). If $d_2=d_0\backslash d_1$, we then need to show that

$$\left|\sum_{i\in d_2}\left\{f(t_i)\ell(I_i\cap J)-\int_{I_i\cap J} f\right\}\right|\le 2\varepsilon.$$

[Note $\{(t_i,I_i\cap J)\}$ is not γ-fine if $i\in d_2$ since $t_i\notin I_i\cap J$.] For each $i\in d_2$, $I_i\cap J$ can be written as the difference of 2 intervals whose closures A_i and B_i each contain t_i. [For example, if $J = [a,b]$, $I_i = [a_i, b_i]$ and $a_i \le t_i < a < b_i \le b$, then

$$J \cap I_i = [a, b_i] = [a_i, b_i] \setminus [a_i, a), A_i = [a_i, b_i], B_i = [a_i, a]\,;$$

the other cases are similar.] Then $\{(t_i, A_i) : i \in d_2\}$ and $\{(t_i, B_i) : i \in d_2\}$ are both γ-fine. By Henstock's Lemma, we have

$$\begin{aligned}
&\left|\sum_{i\in d_2}\left\{f(t_i)\ell(I_i \cap J) - \int_{I_i\cap J} f\right\}\right| \\
&\quad = \left|\sum_{i\in d_2}\left\{f(t_i)\ell(A_i \cap J) + f(t_i)\ell(B_i \cap J) - \int_{A_i\cap J} f - \int_{B_i\cap J} f\right\}\right| \\
&\quad \le \left|\sum_{i\in d_2}\left\{f(t_i)\ell(A_i \cap J) - \int_{A_i\cap J} f\right\}\right| \\
&\qquad + \left|\sum_{i\in d_2}\left\{f(t_i)\ell(B_i \cap J) - \int_{B_i\cap J} f\right\}\right| \le \varepsilon + \varepsilon = 2\varepsilon
\end{aligned}$$

as desired.

The second inequality in Lemma 8 is obtained as above by using the second inequality in Henstock's Lemma (3.1).

The term "uniform" is used in Lemma 10 because the estimates hold uniformly for any subinterval J of I.

Although the proof above is phrased in $\mathbb{R}$, the result also holds in $\mathbb{R}^n$ (see [LCL]).

Lewis and Shisha have related improper Riemann integrals to gauge integrals where the function δ generating the gauge as described in Chapter 1 satisfies special conditions. For example, it is shown that a function $f : [a,b] \to \mathbb{R}$ with a singularity at a is improper Riemann integrable on $[a,b]$ if and only if f is gauge integrable with respect to a function δ which is increasing on $(a,b]$. Functions which are integrable with respect to linear functions δ are also considered. See [LS] for details.

Exercises

1. Using the notation of Lemma 1 show that

$$\left|\sum_{i=1}^{n}\left\{|f(t_i)\ell(I_i)| - \left|\int_{I_i} f\right|\right\}\right| \le 2\varepsilon\,.$$

2. Can the condition $\int_a^c f = 0$ for all $c \in [a, b]$ in Corollary 3 be replaced by $\int_a^b f = 0$?
3. Show that $B(1/2, 1/2) = \pi$. (Hint: Try $t = \cos^2 u$.)
4. Show that $B(x, y) = B(y, x)$.
5. Let $f : [a, b] \to \mathbb{R}$ be bounded on $[a, b]$ and integrable over $[c, b]$ for every $a < c \leq b$. Show that f is integrable over $[a, b]$.
6. Let f be positive and integrable on $[0, b]$. If f is decreasing on $(0, b]$, show that $\lim_{x \to 0^+} xf(x) = 0$.
7. Determine which of the following integrals exist:

$$\text{(a)} \int_0^1 \sqrt{x}/\ln x dx\,, \quad \text{(b)} \int_2^3 \frac{x^2+1}{x^2-4} dx\,,$$

$$\text{(c)} \int_0^1 \frac{dx}{\sqrt{x+x^2}}\,, \quad \text{(d)} \int_0^1 \frac{\ln x}{1-x^2} dx\,.$$

8. Use Example 6 to show that the product of integrable functions needn't be integrable.
9. Let $a_k \in \mathbb{R}$ for $k \in \mathbb{N}$ and define $f : [0, 1] \to \mathbb{R}$ by $f(0) = 0$, $f(t) = 2^k a_k$ for $1/2^k < t \leq 1/2^{k-1}$. Show that f is integrable if and only if $\sum_{k=1}^{\infty} a_k$ converges and in this case $\int_0^1 f = \sum_{k=1}^{\infty} a_k$. Give necessary and sufficient conditions for f to be absolutely integrable. [Hint: Use Theorem 4. A direct proof is possible but complicated; see [Go1] p. 320.]

Chapter 4

The Gauge Integral over Unbounded Intervals

In this chapter we extend the definition of the gauge integral to unbounded intervals in $\mathbb{R}$. This definition could have been made directly in Chapter 1, but it is best to begin the study of the gauge integral on closed, bounded intervals as was carried out in Chapters 1–3. Suppose I is any interval in $\mathbb{R}$, $f : I \to \mathbb{R}$ and we wish to define the integral of f over I. We can extend the definition of f to $\mathbb{R}$ by setting $f(t) = 0$ for $t \in \mathbb{R}\backslash I$ and then the integral of this extension over $\mathbb{R}$ would give a definition of the integral of f over I. Thus, we need only consider the definition of the integral for functions $f : \mathbb{R} \to \mathbb{R}$.

If we wish to extend the definition of the integral for functions $f : \mathbb{R} \to \mathbb{R}$, the first thing that we must consider are the meanings of partitions and tagged partitions of $\mathbb{R}$. A *partition* of $\mathbb{R}$ is a finite number of non-overlapping, closed subintervals of $\mathbb{R}$ whose union is $\mathbb{R}$; of course, in any such partition at least one of the subintervals must have infinite length. A *tagged partition* of $\mathbb{R}$ is a finite collection of pairs $\{(t_i, I_i) : 1 \le i \le m\}$, where $\{I_i : 1 \le i \le m\}$ is a partition of $\mathbb{R}$ and $t_i \in I_i$ for each i. If $f : \mathbb{R} \to \mathbb{R}$ is a strictly positive function and (t_i, I_i) is an element of a tagged partition where I_i has infinite length, then the term $f(t_i)\ell(I_i)$ would be infinite and the corresponding Riemann sum for this tagged partition would also be infinite. If we associate the integral of a positive function with the area under the graph of this function, this is clearly an undesirable situation.

We can easily remedy this situation by considering the extended real numbers. We extend $\mathbb{R}$ by adding the two points at infinity, $\pm\infty$. We denote this extension by $\mathbb{R}^*$ and adopt the following order and algebraic properties in $\mathbb{R}^*$:

$$-\infty < x < \infty \text{ for every } x \in \mathbb{R}\,,$$

$$\infty + \infty = \infty\,,$$

$$-\infty - \infty = -\infty\,,$$

$$x \pm \infty = \pm\infty \text{ for } x \in \mathbb{R}\,,$$

$$x \cdot (\pm\infty) = \pm\infty \text{ for } x > 0\,,$$

$$x \cdot (\pm\infty) = \mp\infty \text{ for } x < 0\,.$$

The arithmetic definitions in $\mathbb{R}^*$ are motivated by the desire to carry forth the properties of limits to functions or sequences with limits in $\mathbb{R}^*$. We now also make the somewhat unorthodox definition that

$$0 \cdot (\pm\infty) = 0\,;$$

this definition is motivated by the integration theory which we are about to develop.

We refer to intervals of the form $[-\infty, a]$, $[a, \infty]$, $[-\infty, \infty]$ as closed subintervals of $\mathbb{R}^*$ and, similarly, call intervals of the form $[-\infty, a)$, $(a, \infty]$ open subintervals of $\mathbb{R}^*$. For unbounded intervals I of the form above, we set $\ell(I) = \infty$.

Let I be a closed subinterval of $\mathbb{R}^*$. A *partition* of I is a finite collection of closed, non-overlapping subintervals of I whose union is I, and a tagged partition of I is a finite collection of ordered pairs $\mathcal{D} = \{(t_i, I_i) : 1 \le i \le m\}$ such that $\{I_i : 1 \le i \le m\}$ is a partition of I and $t_i \in I_i$ for each i. As in Chapter 1 we refer to the I_i as the subintervals of $\mathcal{D}$ and call t_i the *tag* associated with I_i.

If I is any subinterval of $\mathbb{R}$ and $f : I \to \mathbb{R}$, we always assume that f is extended to $\mathbb{R}^*$ by setting $f(t) = 0$ for $t \in \mathbb{R}^* \backslash I$ and that $f(\pm\infty) = 0$ for any function. If I is a closed subinterval of $\mathbb{R}^*$, $f : I \to \mathbb{R}$ and $\mathcal{D} = \{(t_i, I_i) : 1 \le i \le m\}$ is a tagged partition of I, the *Riemann sum* of f with respect to $\mathcal{D}$ is defined to be

$$S(f, \mathcal{D}) = \sum_{i=1}^{m} f(t_i)\ell(I_i)\,.$$

Note that if $\pm\infty$ is the tag associated with the intervals of infinite length, then $S(f, \mathcal{D})$ is always well-defined and is a real number because of our agreement that $0 \cdot (\pm\infty) = 0$.

A *gauge* γ on $I \subset \mathbb{R}^*$ is a function defined on I such that $\gamma(t)$ is an open interval containing t. If $\mathcal{D} = \{(t_i, I_i) : 1 \le i \le m\}$ is a tagged partition of I and

γ is a gauge on I, we say that $\mathcal{D}$ is γ-fine, written $\mathcal{D} << \gamma$, if $t_i \in I_i \subset \gamma(t_i)$ for $1 \le i \le m$. Note that if $\gamma(t)$ is a bounded open interval for every $t \in \mathbb{R} \cap I$, then the tag of any unbounded interval in a γ-fine tagged partition $\mathcal{D}$ of I must be $\pm\infty$ so the Riemann sum of any function $f : I \to \mathbb{R}$ with respect to $\mathcal{D}$ is well-defined.

We are now in a position to define the integral over an arbitrary closed subinterval I of $\mathbb{R}^*$. First we must extend Theorem 1.3 to unbounded intervals.

Theorem 1. *Let I be a closed interval in $\mathbb{R}^*$ and γ a gauge on I. Then there exists a γ-fine tagged partition of I.*

Proof: Consider the case $I = [a, \infty]$. Let $\gamma(\infty) = (b, \infty]$. If $b < a$, set $\mathcal{D} = \{(\infty, I)\}$. If $b \ge a$, there is a γ-fine tagged partition $\mathcal{D}_0$ of $[a, b+1]$. Then $\mathcal{D} = \mathcal{D}_0 \cup \{(\infty, [b+1, \infty])\}$ is a γ-fine tagged partition of I.
The other cases of unbounded intervals are similar.

We now give the definition of the integral over an arbitrary closed subinterval I of $\mathbb{R}^*$. (Exercises 1 and 2 and give other approaches to defining the integral over unbounded intervals.)

Definition 2. *Let I be a closed interval in $\mathbb{R}^*$ and $f : I \to \mathbb{R}$. Then f is (gauge) integrable over I if there exists $A \in \mathbb{R}$ such that for every $\varepsilon > 0$ there is a gauge γ on I such that $|S(f, \mathcal{D}) - A| < \varepsilon$ whenever $\mathcal{D}$ is a γ-fine tagged partition of I.*

As in Theorem 1.4 it is easily established that the value A in Definition 2 is unique (and well-defined by Theorem 1) so we adopt the notation $A = \int_I f = \int_I f(t)dt$. If $I = [a, \infty] (I = [-\infty, a], I = [-\infty, \infty])$, we sometimes write $\int_I f = \int_a^\infty f (\int_{-\infty}^a f, \int_{-\infty}^\infty f)$.

We give an example illustrating the integral in Definition 2.

Example 3. *Let $f(t) = 1/t^2$ for $t \ge 1$. We claim that f is integrable over $[1, \infty)$ with $\int_1^\infty f = 1$. Let $\varepsilon > 0$. We will define a gauge γ on $[1, \infty]$ such that $\gamma(t)$ is bounded for every $t \in [1, \infty)$; this will force the tag of an unbounded interval in any tagged partition to be $+\infty$. If we define $\gamma(\infty) = (2/\varepsilon, \infty]$ and we have $[w, \infty] \subset \gamma(\infty)$, then $|f(\infty)\ell([w, \infty)) - 1/w| = 1/w < \varepsilon/2$. (Note the area under the graph of f over the interval $[w, \infty)$ should be $1/w$.) Next, if $(z, [u, v])$ is a pair in a tagged partition, we want to estimate $|f(z)(v-u) - (1/u - 1/v)|$. (Again the area under the graph of f over $[u, v]$ should be $1/u - 1/v$.) We have*

$$|f(z)(v-u) - (1/u - 1/v)| = \left|\frac{uv - z^2}{z^2}\right| |1/u - 1/v| \le \frac{v-u}{z}(1/v - 1/u)$$

by considering the cases when $uv - z^2$ *is positive or negative. Suppose now that*

$$\mathcal{D} = \{(t_i, I_i) : 1 \le i \le m\}$$

is a tagged partition of $[1, \infty]$ *with* $I_i = [x_{i-1}, x_i], x_0 = 1, x_m = \infty, t_m = \infty$. *From the inequalities above and the fact that*

$$-1 + \frac{1}{x_m} = \sum_{i=1}^{m-1} \left(-\frac{1}{x_{i-1}} + \frac{1}{x_i} \right),$$

we have

$$\begin{aligned} |S(f, \mathcal{D}) - 1| &\le |f(\infty)\ell(I_m) - 1/x_{m-1}| \\ &\quad + \left| \sum_{i=1}^{m-1} \frac{1}{t_i^2}(x_i - x_{i-1}) - 1 + 1/x_{m-1} \right| \\ &\le \varepsilon/2 + \sum_{i=1}^{m-1} \frac{x_i - x_{i-1}}{t_i}(1/x_{i-1} - 1/x_i)\,. \end{aligned}$$

If we define $\gamma(t) = (t - \varepsilon t/4, t + \varepsilon t/4)$ *and assume that* $\mathcal{D} << \gamma$, *we have*

$$\begin{aligned} |S(f, \mathcal{D}) - 1| &\le \varepsilon/2 + \sum_{i=1}^{m-1} \frac{\varepsilon t_i}{2t_i}(1/x_{i-1} - 1/x_i) \\ &= \varepsilon/2 + \varepsilon/2 \sum_{i=1}^{m-1} (1/x_{i-1} - 1/x_i) \\ &= \varepsilon/2 + \varepsilon/2(1 - 1/x_{m-1}) < \varepsilon \end{aligned}$$

and the claim above is justified. Note that the gauge γ *defined above definitely has intervals of variable length with the length being quite large for t large (consider the graph of* f*).*

The basic properties of the integral established in 2.1-2, 2.4-7, 2.9 and particularly Henstock's Lemma 3.1 carry forward to the integral over unbounded intervals. The reader is invited to review the proofs of these results.

As was the case for bounded intervals, we next show that there are no "improper integrals" over unbounded intervals.

Theorem 4. *Let $f : [a,\infty] = I \to \mathbb{R}$ be integrable over $[a,b]$ for every $a < b < \infty$. Then f is integrable over I if and only if $\lim_{b\to\infty} \int_a^b f = A$ exists. In the case, $A = \int_I f$.*

Proof: Let $\varepsilon > 0$. Suppose f is integrable over I. There is a gauge γ on I such that $|S(f,\mathcal{D}) - \int_I f| < \varepsilon/2$ whenever $\mathcal{D} << \gamma$. For $a < c < \infty$, there is a gauge γ_c on $[a,c]$ such that $|S(f,\mathcal{E}) - \int_a^c f| < \varepsilon/2$ whenever $\mathcal{E}$ is a γ-fine tagged partition of $[a,c]$. We may assume that $\gamma_c(t) \subset \gamma(t)$ for every $t \in [a,c]$. Let $\gamma(\infty) = (T,\infty]$. For $c > T$, let $\mathcal{E}$ be a γ_c-fine tagged partition of $[a,c]$. Set $\mathcal{D} = \mathcal{E} \cup \{(\infty,[c,\infty])\}$. Then $\mathcal{D} << \gamma$ and

$$\left|\int_I f - \int_a^c f\right| \le \left|\int_I f - S(f,\mathcal{D})\right| + \left|S(f,\mathcal{E}) - \int_a^c f\right|$$

$$+ |f(\infty)|\ell([c,\infty]) < \varepsilon/2 + \varepsilon/2 = \varepsilon\,.$$

Hence, $\lim_{c\to\infty} \int_a^c f = \int_I f$.

Suppose $\lim_{c\to\infty} \int_a^c f = A$. Choose a sequence $\{c_k\}$ such that $a = c_0 < c_1 < c_2 < \ldots$ and $c_k \uparrow \infty$. Choose a gauge γ_0 on $[c_0,c_1]$ such that $|S(f,\mathcal{D}) - \int_{c_0}^{c_1} f|$ $< \varepsilon/2^2$ whenever $\mathcal{D}$ is a γ_0-fine tagged partition of $[c_0,c_1]$. For $k \ge 1$, choose a gauge γ_k on $[c_{k-1},c_{k+1}]$ such that $|S(f,\mathcal{D}) - \int_{c_{k-1}}^{c_{k+1}} f| < \varepsilon/2^{k+2}$ whenever $\mathcal{D}$ is a γ_k-fine tagged partition of $[c_{k-1},c_{k+1}]$.

Choose N such that $b \ge c_N$ implies $|\int_a^b f - A| < \varepsilon/2$. Now define a gauge γ on I by

$$\gamma(t) = \begin{cases} (c_N,\infty] & t = \infty \\ \gamma_k(t) \cap (c_{k-1},c_{k+1}) & c_k \le t < c_{k+1} \qquad k = 1,2,\ldots \\ \gamma_0(t) \cap (c_0 - 1, c_1) & c_0 \le t < c_1\,. \end{cases}$$

Suppose that $\mathcal{D} = \{(t_i,I_i) : 1 \le i \le m\}$ is a γ-fine tagged partition of I. Assume that $I_m = [\alpha,\infty]$ is the unbounded interval in $\mathcal{D}$; note $t_m = \infty$ and $c_N < \alpha$. For $k \ge 0$ let $\mathcal{D}_k$ be the elements in $\mathcal{D}$ whose tags are in $[c_k,c_{k+1}]$. By the definition of γ each $\mathcal{D}_k$ is γ_k-fine; if J_k is the union of the subintervals in $\mathcal{D}_k$, then $J_0 \subset [c_0,c_1)$ and $J_k \subset [c_{k-1},c_{k+1})$ for $k \ge 1$. By Henstock's Lemma (3.1)

$$\left|S(f,\mathcal{D}_k) - \int_{J_k} f\right| \le \varepsilon/2^{k+2}$$

so

$$
\begin{aligned}
|A - S(f,\mathcal{D})| &\le \left|A - \int_a^\alpha f\right| + \left|\int_a^\alpha f - S(f,\mathcal{D})\right| \\
&< \varepsilon/2 + \left|\sum_{k=0}^{\infty} \int_{J_k} f - \sum_{k=0}^{\infty} S(f,\mathcal{D}_k) + f(\infty)\ell(I_m)\right| \\
&\le \varepsilon/2 + \sum_{k=0}^{\infty} \varepsilon/2^{k+2} = \varepsilon\,.
\end{aligned}
$$

Of course, an analogous result holds for integrals over intervals of the form $[-\infty, b], [-\infty, \infty]$.

Using Theorem 4, we can obtain an analogue of Corollary 3.5 for unbounded intervals.

Corollary 5. *Let $f : [a, \infty] = I \to \mathbb{R}$ be absolutely integrable over $[a, b]$ for every $a < b < \infty$.*

(i) *If f is non-negative, f is integrable over I if and only if*

$$
\sup\left\{\int_a^b f : a < b < \infty\right\} < \infty\,.
$$

(ii) *If $g : I \to \mathbb{R}$ is non-negative and integrable over I and $|f(t)| \le g(t)$ for $t \in I$, then f is absolutely integrable over I.*

Using Corollary 4.11 below it will follow that the absolute integrability assumption in Corollary 5 can be relaxed to integrability.

Example 6. *Let $p \in \mathbb{R}$ and set $f(t) = 1/t^p$ for $t \neq 0$. Then $\int_1^b f = (b^{-p+1} - 1)/(1-p)$ for $p \neq 1$; if $p = 1$, $\int_1^b f = \ln b$. Thus, from Corollary 5, it follows that f is integrable over $[1, \infty]$ if and only if $p > 1$. [Compare with Example 3.6.]*

We can employ Corollary 5 to establish the integral test for convergence of series.

Proposition 7. *Let $f : [1, \infty] \to \mathbb{R}$ be positive, decreasing and integrable over $[1, b]$ for every $1 < b < \infty$. Then the integral $\int_1^\infty f = A$ exists if and only if the series $\sum_{k=1}^\infty f(k) = S$ converges. In this case, $A \le S \le A + f(1)$.*

Proof: For $i \le x \le i + 1$, $f(i+1) \le f(x) \le f(i)$. Thus,

$$
f(i+1) \le \int_i^{i+1} f \le f(i)
$$

and

$$\sum_{i=1}^{n-1} f(i+1) \le \int_1^n f \le \sum_{i=1}^{n-1} f(i).$$

Letting $n \to \infty$ gives the first statement and, in the case of convergence, yields

$$S - f(1) \le A \le S.$$

We give another example of a conditionally convergent integral. [As before an integrable function f is absolutely (conditionally) integrable if $|f|$ is (not) integrable.]

Example 8. *The integral* $\int_1^\infty \frac{\sin t}{t} dt$ *exists. In fact, integration by parts gives*

$$\int_1^b \frac{\sin t}{t} dt = -\frac{\cos t}{t}\bigg|_1^b - \int_1^b \frac{\cos t}{t^2} dt.$$

The function $(\cos t)/t^2$ *is integrable over* $[1, \infty]$ *by Corollary* 5 (ii) *and Example* 7 (*with* $p = 2$). *Since* $\lim_{b\to\infty}(\cos b)/b = 0$, *it follows from Theorem* 4 *that the integral of* $(\sin t)/t$ *over* $[1, \infty]$ *exists.*

However, $\int_1^\infty |\frac{\sin t}{t}| dt$ *does not exist. Indeed, we have*

$$\begin{aligned}\int_\pi^{n\pi} \left|\frac{\sin t}{t}\right| dt &= \sum_{k=1}^{n-1} \int_{k\pi}^{(k+1)\pi} \left|\frac{\sin t}{t}\right| dt \\ &\ge \sum_{k=1}^{n-1} \frac{1}{(k+1)\pi} \int_{k\pi}^{(k+1)\pi} |\sin t| dt \\ &= \sum_{k=1}^{n-1} \frac{2}{(k+1)\pi}\end{aligned}$$

for all n. [*See also Exercise* 8.17.]

We next consider absolute integrability for the gauge integral. Recall that the gauge integral admits conditionally integrable functions (Examples 8, 2.12 or Exercise 3.9). We now give necessary and sufficient conditions for absolute integrability. These conditions involve the variation of a function; for those readers unfamiliar with the definitions and properties of functions of bounded variation, they are given in Appendix 2. We begin with the case of functions defined on a closed bounded interval.

Theorem 9. *Let $I = [a, b]$ with $-\infty < a < b < \infty$ and $f : I \to \mathbb{R}$ be integrable over I. Let $F(x) = \int_a^x f, a \le x \le b$, be the indefinite integral of f. Then $|f|$ is integrable over I if and only if F is of bounded variation over I. In this case, $\int_I |f| = Var(F : I)$.*

Proof: Let $V = Var(F : I)$. If $|f|$ is integrable and $a = x_0 < x_1 < \cdots < x_n = b$ is a partition of I, then

$$\sum_{i=1}^{n} |F(x_i) - F(x_{i-1})| \le \sum_{i=1}^{n} \int_{x_{i-1}}^{x_i} |f| = \int_a^b |f|$$

so $V \le \int_a^b |f| < \infty$.

For the converse assume $V < \infty$ and let $\varepsilon > 0$. There is a partition

$$\mathcal{P} = \{a = x_0 < x_1 < \cdots < x_n = b\}$$

of I such that $V - \varepsilon < \sum_{i=1}^{n} |\int_{K_i} f| \le V$, where $K_i = [x_{i-1}, x_i]$. Note that if $\mathcal{P}_1 = \{a = y_0 < y_1 < \cdots < y_m = b\}$ is a refinement of the partition $\mathcal{P}$(*i.e.*, $\mathcal{P}_1 \supset \mathcal{P}$) and $L_i = [y_{i-1}, y_i]$, then

$$(1) \quad V - \varepsilon \le \sum_{i=1}^{n} \left| \int_{K_i} f \right| \le \sum_{i=1}^{m} \left| \int_{L_i} f \right| \le V$$

(Appendix 2.)

Let γ_1 be a gauge on I such that $|S(f, \mathcal{D}) - \int_I f| < \varepsilon$ whenever $\mathcal{D}$ is a γ_1-fine tagged partition of I. By Exercise 3.1 if $\mathcal{D} = \{(t_i, J_i) : 1 \le i \le p\} << \gamma_1$, then

$$(2) \quad \left| \sum_{i=1}^{p} \left\{ |f(t_i)| \ell(J_i) - \left| \int_{J_i} f \right| \right\} \right| \le 2\varepsilon.$$

Let γ be a gauge on I such that $\gamma(t) \subset \gamma_1(t)$ for every $t \in I, \gamma(t) \subset K_i$ when $t \in K_i^o$ and $\gamma(x_i) \subset (x_{i-1}, x_{i+1})$ where $x_{-1} = -\infty$, $x_{n+1} = \infty$. If $\mathcal{E} = \{(z_i, I_i) : 1 \le i \le q\}$ is a γ-fine tagged partition of I, then there is a γ-fine tagged partition $\mathcal{E}' = \{(z_i', I_i') : 1 \le i \le r\}$ such that the partition $\{I_i' : 1 \le i \le r\}$ is a refinement of $\{K_1, \ldots, K_n\}$ and such that $S(|f|, \mathcal{E}) = S(|f|, \mathcal{E}')$. In fact, by the definition of γ, we may take

$$\mathcal{E}' = \{(z_i, I_i \cap K_j) : 1 \le i \le q, 1 \le j \le n, I_i^o \cap K_j^o \ne \emptyset\}$$

(Figure 4.1). Then (1) and (2) give

$$V - \varepsilon \le \sum_{i=1}^{r} \left| \int_{I_i'} f \right| \le V \text{ and } \left| \sum_{i=1}^{r} \left\{ |f(z_i')| \ell(I_i') - \left| \int_{I_i'} f \right| \right\} \right| \le 2\varepsilon$$

so

$$|S(|f|, \mathcal{E}) - V| = \left| S(|f|, \mathcal{E}') \pm \sum_{i=1}^{r} \left| \int_{I_i'} f \right| - V \right| \le 2\varepsilon + \varepsilon$$

and the result follows.

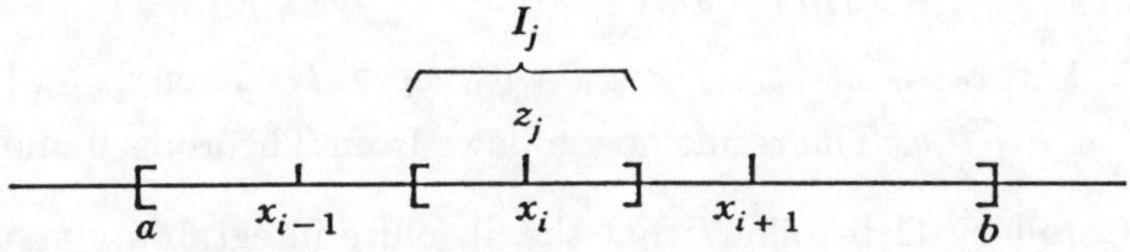

Figure 4.1

We next consider the case when the interval I is unbounded.

Theorem 10. *Let $I = [-\infty, b]$, $b < \infty$, and $f : I \to \mathbb{R}$ be integrable over I. Let $F(x) = \int_{-\infty}^{x} f$ for $-\infty < x \le b$. Then $|f|$ is integrable over I if and only if F is of bounded variation over I. In this case, $\int_I |f| = Var(F : I)$.*

Proof: For $-\infty < a < b$, set $F_a(x) = \int_a^x f$ for $a \le x \le b$ and $c_a = \int_{-\infty}^{a} f$ so $F(x) = F_a(x) + c_a$ for $a \le x \le b$.

Suppose $|f|$ is integrable over I. If $-\infty < a < b$, then by Theorem 9

$$\int_a^b |f| = Var(F_a : [a, b]) = Var(F : [a, b]).$$

By Theorem 4

$$\int_{-\infty}^{b} |f| = \lim_{a \to -\infty} \int_a^b |f| = Var(F : I).$$

Suppose F is of bounded variation over I. If $-\infty < a < b$, then by Theorem 9

$$\int_a^b |f| = Var(F_a : [a, b]) = Var(F : [a, b]).$$

By Theorem 4

$$Var(F:I) = \lim_{a\to-\infty} Var(F:[a,b]) = \lim_{a\to-\infty}\int_a^b |f| = \int_{-\infty}^b |f| .$$

The cases when $I = [a,\infty]$ or $[-\infty,\infty]$ are treated similarly.

From Theorems 9 and 10 we have the following comparison theorem for absolute integrability.

Corollary 11. *Let I be a closed subinterval of $\mathbb{R}^*$ and let $f, g : I \to \mathbb{R}$ be integrable over I with $|f(t)| \le g(t)$ for every $t \in I$. Then f is absolutely integrable over I with $|\int_I f| \le \int_I |f| \le \int_I g$.*

Proof: Let $x_0 < x_1 < \cdots < x_n$ with $x_i \in I$. Then $\sum_{i=1}^n |\int_{x_{i-1}}^{x_i} f| \le \sum_{i=1}^n \int_{x_{i-1}}^{x_i} g = \int_{x_0}^{x_n} g$. The result now follows from Theorems 9 and 10.

From Corollary 11 it follows that the absolute integrability assumption in Corollaries 3.5 and 4.5 can be relaxed to integrability.

If $f, g : I \to \mathbb{R}$, we write $f \vee g = \max(f,g)$, $f \wedge g = \min(f,g)$. In contrast to the situation which obtains for both the Riemann and Lebesgue integrals, if f and g are (gauge) integrable, $f \vee g$ and $f \wedge g$ may fail to be integrable. Indeed, if $f^+ = f \vee 0$ and $f^- = (-f) \vee 0$, $|f| = f^+ + f^-$, $f^+ = (f+|f|)/2$ and $f^- = (-f+|f|)/2$ so $|f|$ is integrable if and only if both f^+ and f^- are integrable. Thus if $f(t) = (\sin t)/t$ for $t \in I = [1,\infty]$, then by Example 8 both f^+ and f^- are not integrable over I.

We do have sufficient conditions for the max and min of integrable functions to be integrable.

Proposition 12. *Let I be a closed subinterval of $\mathbb{R}^*$ and let $f, g, h : I \to \mathbb{R}$ be integrable over I.*

(i) *If f and g are absolutely integrable over I, then $f \vee g$ and $f \wedge g$ are integrable over I.*
(ii) *If $f \le h$ and $g \le h$, then $f \vee g$ and $f \wedge g$ are integrable over I.*
(iii) *If $h \le f$ and $h \le g$, then $f \vee g$ and $f \wedge g$ are integrable over I.*

Proof: (i) follows from the identities

$$f \vee g = [f+g+|f-g|]/2 \text{ and } f \wedge g = [f+g-|f-g|]/2 ,$$

the inequality $|f-g| \le |f| + |g|$ and Corollary 11.

To prove (ii) note that $f \vee g \leq h$ and, therefore,

$$|f - g| = 2f \vee g - f - g \leq 2h - f - g\,.$$

Thus, $|f - g|$ is integrable by Corollary 11. The result now follows from the identities for $f \vee g$ and $f \wedge g$ given above.

Part (iii) is established in the same way using $|f - g| \leq f + g - 2h$.

We conclude this section by considering the integrability of the product of 2 integrable functions. We first establish a lemma which gives a sufficient condition for the integrability of a product; this lemma is sometimes called Dedekind's Test.

Lemma 13. *Let $f, g : [a, b] = I \to \mathbb{R}$ be continuous on $(a, b]$ with $\lim_{x \to a^+} g(x) = 0$ and g' absolutely integrable over I. Assume that $F(x) = \int_x^b f, a < x \leq b$, is bounded on $(a, b]$. Then fg is integrable over I.*

Proof: For $a < c \leq b$, we have $(Fg)' = -fg + Fg'$ so Fg' is integrable over $[c, b]$, and since F is bounded on $(a, b]$ and g' is absolutely integrable, Fg' is absolutely integrable over $[a, b]$ by Corollary 11 and 3.5. We have $\int_c^b fg = \int_c^b Fg' + F(c)g(c)$, and since F is bounded and $\lim_{x \to a^+} g(x) = 0$, fg is integrable over $[a, b]$ with $\int_a^b fg = \int_a^b Fg'$ by Theorem 3.4.

We now use Dedekind's Test to give an example illustrating a multiplicative property (lack-there-of) for the gauge integral.

Example 14. *Let $f(t) = t^{-2} \cos(\frac{1}{t}) = (-\sin(\frac{1}{t}))'$ and $g(t) = t^{1/2}$ for $0 \leq t \leq 1 (f(0) = 0)$. By Example 3.6 f and g satisfy the hypotheses of Lemma 13 so $h(t) = t^{-3/2} \cos(\frac{1}{t})$ is integrable over $[0, 1]$. Let $m(t) = \cos(\frac{1}{t})$. Then m is integrable over $[0, 1]$ (Exercise 16), but mh is not integrable over $[0, 1]$ (Exercise 15).*

This example shows that in general the product of an integrable function and a bounded, integrable function need not be integrable. (See the Notes/Remarks section for further results and references.)

Notes/Remarks

Two alternative approaches to defining the gauge integral over unbounded intervals are given in Exercises 1 and 2; Exercise 1 is from [LPY] and Exercise 2 from [M].

Lewis and Shisha have connected the gauge integral over the interval $[0, \infty]$ with what they call a "simple integral"; the simple integral essentially uses a constant length gauge and so is an analogue of the classical Riemann integral ([LS]). They also show that a function has an improper Riemann or Cauchy-Riemann integral if and only if the function is integrable with respect to a monotone function δ which generates the gauge as in Chapter 1.

Example 14 motivates the following "multiplier problem" for the gauge integral. Characterize those functions $g : [a, b] \to \mathbb{R}$ such that fg is gauge integrable for every gauge integrable function f; let us refer to each such function g as a *multiplier*. Example 14 shows that even a bounded, integrable function may not be a multiplier. We show that a function of bounded variation is always a multiplier and then state a partial converse to this result. The proof uses the Riemann-Stieltjes integral (see [Ru], Chapter 6 for a development); we give a reference to the result employed. Also, employed in the proof is Abel's "partial summation formula":

$$\sum_{k=1}^{n} a_k b_k = a_n B_n + \sum_{k=1}^{n-1} B_k(a_k - a_{k+1}), \text{ where } B_k = \sum_{i=1}^{k} b_i \, .$$

Theorem 15*. *Let $f : [a, b] = I \to \mathbb{R}$ be integrable over I and set $F(t) = \int_a^t f$. Let $g : I \to \mathbb{R}$ be of bounded variation. Then fg is integrable over I with $\int_a^b fg = F(b)g(b) - \int_a^b F dg$ [this formula is sometimes referred to as "integration by parts"].*

Proof: Let $\varepsilon > 0$. There is a gauge γ on I such that $|S(f, \mathcal{D}) - \int_a^b f| < \varepsilon$ whenever $\mathcal{D} << \gamma$. Since the Stieltjes integral of F with respect to g exists, there is a $\delta > 0$ such that

$$\left| \sum_{i=1}^{m} F(t_i)(g(x_i) - g(x_{i-1})) - \int_I F dg \right| < \varepsilon$$

whenever $a = x_0 < x_1 < \cdots < x_m = b$ is a partition of $[a, b]$ with $x_i - x_{i-1} < \delta$ and $x_{i-1} \le t_i \le x_i$ for every i (Appendix 2, Theorem 8 and [Ru] 6.14). Define a new gauge γ' by requiring that $\gamma'(t) \subset \gamma(t) \cap (t - \delta, t + \delta)$ and $\gamma'(t) \subset \gamma(t) \cap (a, b)$ when $t \in (a, b)$. Suppose that $\mathcal{D} = \{(t_i, I_i) : 1 \le i \le n\}$ is γ'-fine with $I_i = [x_{i-1}, x_i]$, $a = x_0 < x_1 < \cdots < x_n = b$; note $x_i - x_{i-1} < \delta$, $t_i = a$ and $t_n = b$ so $a = t_0 < t_1 < \cdots < t_n = b$ is a partition of $[a, b]$. Using Abel's partial summation formula with $a_k = g(t_k)$, $b_k = f(t_k)(x_k - x_{k-1})$, and setting

$A = F(b)g(b) - \int_a^b Fdg$, we have from Henstock's Lemma 3.1

$$\begin{aligned}
&\left|\sum_{k=1}^{n} f(t_k)g(t_k)(x_k - x_{k-1}) - A\right| \\
&= \left|\sum_{k=1}^{n-1}\left[\sum_{i=1}^{k} f(t_i)(x_i - x_{i-1})\right][g(t_k) - g(t_{k+1})]\right. \\
&\qquad \left. + \left[\sum_{i=1}^{n} f(t_i)(x_i - x_{i-1})\right] g(t_n) - A\right| \\
&\le \left|\sum_{k=1}^{n-1}\left[\sum_{i=1}^{k} f(t_i)(x_i - x_{i-1}) - F(x_k)\right][g(t_k) - g(t_{k+1})]\right| \\
&\qquad + \left|\sum_{k=1}^{n-1} F(x_k)[g(t_{k+1}) - g(t_k)] - \int_a^b Fdg\right| \\
&\qquad + \left|\sum_{i=1}^{n-1} f(t_i)(x_i - x_{i-1}) - F(b)\right| |g(b)| \\
&\le \varepsilon Var(g : I) + \varepsilon + |g(b)|\varepsilon\,.
\end{aligned}$$

Hence, the result follows.

There is a partial converse to Theorem 15 which gives a characterization of multipliers. We say that a property P holds *almost everywhere* (abbreviated *a.e.*) in a subset $E \subset \mathbb{R}$ if the property holds for all the points in E except those in a null subset of E (Definition 1.9).

Theorem 16*. *If $g : [a, b] \to \mathbb{R}$ is a multiplier, then g is equal a.e. to a function*

$$g_1 : [a, b] \to \mathbb{R}$$

which is of bounded variation.

See [LPY], Theorem 12.9, for a proof which makes extensive use of Lebesgue measure. For a characterization of multipliers in higher dimensions, see [LCL1].

Exercises 6, 8, 10 and 17 give some further results on the integrability of products of integrable functions.

Exercises

1. Let $f : \mathbb{R} \to \mathbb{R}$. Show that f is integrable over $\mathbb{R}$ if and only if there exists $A \in \mathbb{R}$ such that for every $\varepsilon > 0$ there exist $a < b$ and a gauge γ on $[a, b]$ such that $|S(f, \mathcal{D}) - A| < \varepsilon$ whenever $\mathcal{D}$ is a γ-fine tagged partition of $[a, b]$.
2. Let $f : \mathbb{R} \to \mathbb{R}$. Show that f is integrable over $\mathbb{R}$ if and only if there exists $A \in \mathbb{R}$ such that for every $\varepsilon > 0$ there exist $r > 0$ and a gauge γ on $\mathbb{R}$ such that if $a \leq -r$ and $b \geq r$, then $|S(f, \mathcal{D}) - A| < \varepsilon$ whenever $\mathcal{D}$ is a γ-fine tagged partition of $[a, b]$.
3. Let $a_k \in \mathbb{R}$ for $k \in \mathbb{N}$. Define $f : [1, \infty) \to \mathbb{R}$ by $f(t) = a_k$ for $k \leq t < k+1$. Show that f is integrable over $[1, \infty)$ if and only if $\sum a_k$ converges. Give necessary and sufficient conditions for f to be absolutely integrable.
4. Let $f : [a, \infty) \to \mathbb{R}$ be differentiable. Give necessary and sufficient conditions for f' to be integrable over $[a, \infty)$.
5. Show that $\int_0^\infty \sin(x^2) dx$ exists. (Fresnel Integral). Is the integral absolutely convergent? [Hint: try $t = x^2$.]
6. Let $f, g : [a, b] \to \mathbb{R}$. Suppose fg and f are integrable over $[a, c]$ for every $a \leq c < b$ and $F(t) = \int_a^t f$ for $a \leq t < b$. Assume $\lim_{t \to b^-} F(t)$ exists and g is differentiable with g' absolutely integrable over $[a, b]$. Show that fg is integrable over $[a, b]$. [Hint: Integrate by parts.]
7. (Limit Form of Comparison Test) Let $f, g : [a, b] \to \mathbb{R}$ and assume f and g are integrable over $[a, c]$ for $a \leq c < b$ with $g(t) > 0$ for $a \leq t \leq b$. Let $\lim_{t \to b^-} f(t)/g(t) = L \in \mathbb{R}^*$.

   (i) If $L = 0$ and g is integrable over $[a, b]$, show f is integrable over $[a, b]$.
   (ii) If $0 < L < \infty$, show g is integrable over $[a, b]$ if and only if f is integrable over $[a, b]$.
   (iii) If $L = \infty$ and f is integrable over $[a, b]$, show g is integrable over $[a, b]$.

8. (Abel) Let $f, g : [a, \infty) \to \mathbb{R}$. Suppose f is continuous and set $F(t) = \int_a^t f$. Assume that F is bounded and g is non-negative, differentiable and decreasing. Show that $\int_a^\infty fg$ exists if either (a) $\lim_{t \to \infty} g(t) = 0$ or (b) $\int_a^\infty f$ exists. [Hint: Integrate by parts.]
9. Use Abel's Test in Exercise 8 to show $\int_1^\infty \frac{\sin t}{t^p} dt$ exists for $p > 0$. Show that the integral is conditionally convergent for $0 < p \leq 1$. [Recall Example 8.]
10. Let $f : [a, \infty) \to \mathbb{R}$ be continuous and $F(t) = \int_a^t f$ bounded on $[a, \infty)$. Let $g : [a, \infty) \to \mathbb{R}$ be such that g' is non-positive and continuous on $[a, \infty)$ with

$$\lim_{t \to \infty} g(t) = 0\,.$$

Show that $\int_a^\infty fg$ exists.

11. Use Exercise 10 to show $\int_3^\infty \frac{\sin t}{\log t} dt$ exists.
12. Suppose f is absolutely integrable over I and let $c > 0$. Define $f_c(t) = f(t)$ if $|f(t)| \le c$ and $f_c(t) = 0$ otherwise. Show that f_c is absolutely integrable over I.

13*. Extend the definition of null set given in Definition 1.9 and then extend the results in Exercise 1.9 to $\mathbb{R}$.

14. Let f, g be absolutely integrable over I. Show that $f + g$ is absolutely integrable over I.
15. Use the identity $\cos^2 u = (1 - \cos 2u)/2$ and Example 3.6 to show that the function mh in Example 14 is not integrable.
16. Show that $\cos(\frac{1}{t})$ is integrable over $[0, 1]$. [Hint: Make the change of variable $s = 1/t$ and use Corollary 5 and Example 6.]
17. Let $f, g : [a, b] \to \mathbb{R}$ be continuous on $(a, b]$ and g' absolutely integrable over $[a, b]$. Assume $F(t) = \int_t^b f$ is bounded. Show that fg is integrable over $[a, b]$ if and only if $\lim_{c \to a^+} F(c)g(c)$ exists.

Chapter 5

Convergence Theorems

The principal reason that the Lebesgue integral is favored over the Riemann integral is the fact that convergence theorems of the form $\lim \int_I f_k = \int_I (\lim f_k)$ hold for the Lebesgue integral under very general conditions. The major convergence theorems of this type are the Monotone Convergence Theorem (MCT) and the Dominated Convergence Theorem (DCT). In this section we will show that these same major convergence theorems hold for the gauge integral, showing that the gauge integral enjoys the same advantages over the Riemann integral as the Lebesgue integral.

We begin by introducing the concept which is at the center of the convergence theorems for the gauge integral, uniform integrability. Throughout this section let I be a closed interval (bounded or unbounded) in $\mathbb{R}^*$ and f_k, $f : I \to \mathbb{R}$ for $k \in \mathbb{N}$. Further, let $\mathcal{I}$ be the family of all closed subintervals of I.

Definition 1. *$\{f_k\}$ is uniformly integrable over I if each f_k is integrable over I and if for every $\varepsilon > 0$ there exists a gauge γ on I such that $|S(f_k, \mathcal{D}) - \int_I f_k| < \varepsilon$ for every $k \in \mathbb{N}$ whenever $\mathcal{D} << \gamma$.*

The point of Definition 1 is, of course, that the same gauge works uniformly for all k. For uniformly integrable sequences of integrable functions we have the following convergence theorem.

Theorem 2. *Let $\{f_k\}$ be uniformly integrable over I and assume that $f_k \to f$ pointwise. Then f is integrable over I and $\lim \int_I f_k = \int_I f (= \int_I (\lim f_k))$.*

Proof: Let $\varepsilon > 0$. Let γ be a gauge on I such that $|S(f_k, \mathcal{D}) - \int_I f_k| < \varepsilon/3$ for every k whenever $\mathcal{D} << \gamma$, where we may assume that $\gamma(t)$ is bounded for every $t \in \mathbb{R}$. Fix a γ-fine tagged partition $\mathcal{D}_0$ of I. Pick N such that

$|S(f_i, \mathcal{D}_0) - S(f_j, \mathcal{D}_0)| < \varepsilon/3$ whenever $i, j \geq N$; this is possible by the pointwise convergence of $\{f_k\}$ and the fact that $f_k(\pm\infty) = 0$. If $i, j \geq N$, then

$$\left|\int_I f_i - \int_I f_j\right| \leq \left|\int_I f_i - S(f_i, \mathcal{D}_0)\right| + |S(f_i, \mathcal{D}_0) - S(f_j, \mathcal{D}_0)|$$

$$+ \left|S(f_j, \mathcal{D}_0) - \int_I f_j\right| < \varepsilon\,.$$

Hence, $\lim \int_I f_k = L$ exists.

Suppose now that $\mathcal{D}$ is a γ-fine tagged partition of I. As above pick N such that

$$|S(f_N, \mathcal{D}) - S(f, \mathcal{D})| < \varepsilon/3$$

and also so $|L - \int_I f_N| < \varepsilon/3$. Then

$$|S(f, \mathcal{D}) - L| \leq |S(f, \mathcal{D}) - S(f_N, \mathcal{D})|$$

$$+ \left|S(f_N, \mathcal{D}) - \int_I f_N\right| + \left|\int_I f_N - L\right| < \varepsilon\,.$$

Hence, f is integrable over I with $\int_I f = L = \lim \int_I f_k$ as desired.

We now proceed to establish the major convergence theorems for the gauge integral. Since we are working over arbitrary subintervals of $\mathbb{R}^*$, we need a preliminary lemma which will enable us to treat the case of unbounded intervals.

Lemma 3. *There exists a strictly positive $\varphi : \mathbb{R} \to (0, \infty)$ which is integrable over $\mathbb{R}$ and a gauge $\gamma(= \gamma\varphi)$ such that $0 \leq S(\varphi, \mathcal{D}) \leq 1$ for every γ-fine partial tagged partition $\mathcal{D}$ of $\mathbb{R}$.*

Proof: Pick any strictly positive function φ on $\mathbb{R}$ with $\int_{\mathbb{R}} \varphi = 1/2[\varphi(t) = 1/(2\pi(1+t)^2)]$. There is a gauge γ on $\mathbb{R}$ such that $|S(\varphi, \mathcal{D}) - 1/2| < 1/2$ whenever $\mathcal{D} << \gamma$. Suppose $\mathcal{D} = \{(t_i, I_i) : 1 \leq i \leq m\}$ is a γ-fine partial tagged partition of $\mathbb{R}$ and set $I = \bigcup_{i=1}^m I_i$. By Henstock's Lemma 3.1,

$$\left|S(\varphi, \mathcal{D}) - \int_I \varphi\right| \leq 1/2 \quad \text{so} \quad 0 \leq S(\varphi, \mathcal{D}) \leq 1/2 + \int_I \varphi \leq 1$$

since φ is positive.

We now prove our first convergence result from which the Monotone and Dominated Convergence Theorems will be derived.

Theorem 4. *Let $f_k, f : I \to \mathbb{R}$ be non-negative with each f_k integrable over I and suppose $f = \sum_{k=1}^{\infty} f_k$ pointwise on I with $\sum_{k=1}^{\infty} \int_I f_k < \infty$. If $s_n = \sum_{k=1}^{n} f_k$, then*

(i) *$\{s_n\}$ is uniformly integrable over I,*
(ii) *f is integrable over I and*
(iii) *$\sum_{k=1}^{\infty} \int_I f_k = \int_I f = \int_I \sum_{k=1}^{\infty} f_k$.*

Proof: In order to avoid treating special cases we assume in the proof that $I = \mathbb{R}$; this causes no difficulty since we can always extend functions from I to $\mathbb{R}$ by setting them equal to 0 on $\mathbb{R} \setminus I$. Let $\varepsilon > 0$. For each n pick a gauge γ_n on $\mathbb{R}^*$ with $\gamma_n(t)$ bounded for each $t \in \mathbb{R}$ such that

$$\left| \int_{\mathbb{R}} s_n - S(s_n, \mathcal{D}) \right| < \varepsilon / 2^n$$

whenever $\mathcal{D} << \gamma_n$.

Pick n_0 such that $\sum_{k=n_0}^{\infty} \int_{\mathbb{R}} f_k < \varepsilon$ and for every $t \in \mathbb{R}$ pick $n(t) \geq n_0$ such that $k \geq j \geq n(t)$ implies

$$\left| \sum_{i=j}^{k} f_i(t) \right| < \varepsilon \varphi(t)$$

where φ and $\gamma\varphi$ are as in Lemma 3. Define a gauge γ on $\mathbb{R}^*$ by

$$\gamma(t) = \left(\bigcap_{j=1}^{n(t)} \gamma_j(t) \right) \cap \gamma_\varphi(t)$$

for $t \in \mathbb{R}$ and

$$\gamma(\pm\infty) = \left(\bigcap_{j=1}^{n_0} \gamma_j(\pm\infty) \right) \cap \gamma_\varphi(\pm\infty)$$

and set $n(\pm\infty) = n_0$.

Suppose $\mathcal{D} = \{(t_i, I_i) : 1 \leq i \leq m\} << \gamma$. To establish (i) first note that $\mathcal{D} << \gamma_i$ for $i = 1, \ldots, n_0$ implies $|S(s_i, \mathcal{D}) - \int_{\mathbb{R}} s_i| < \varepsilon/2^i < \varepsilon$. So fix $n > n_0$.

Set

$$d_1 = \{i : 1 \le i \le m, n(t_i) \ge n\}$$

and

$$d_2 = \{i : 1 \le i \le m, n(t_i) < n\},$$

and note $\mathcal{D}_1 = \{(t_i, I_i) : i \in d_1\} << \gamma_n$ by the definition of γ. Set $I = \cup\{I_i : i \in d_1\}$. We have, using Henstock's Lemma 3.1,

$$\begin{aligned}
(1) \left|\int_{\mathbb{R}} s_n - S(s_n, \mathcal{D})\right| &\le \left|\int_I s_n - S(s_n, \mathcal{D}_1)\right| + \left|\sum_{i\in d_2}\sum_{j=1}^{n}\left\{\int_{I_i} f_j - f_j(t_i)\ell(I_i)\right\}\right| \\
&\le \varepsilon/2^n + \left|\sum_{i\in d_2}\sum_{j=1}^{n(t_i)}\left\{\int_{I_i} f_j - f_j(t_i)\ell(I_i)\right\}\right| \\
&\quad + \left|\sum_{i\in d_2}\sum_{j=n(t_i)+1}^{n}\int_{I_i} f_j\right| + \left|\sum_{i\in d_2}\sum_{j=n(t_i)+1}^{n} f_j(t_i)\ell(I_i)\right| \\
&< \varepsilon + T_1 + T_2 + T_3,
\end{aligned}$$

with obvious definitions for T_1, T_2 and T_3.

We first estimate T_3. From Lemma 2,

$$T_3 \le \sum_{i\in d_2}\left|\sum_{j=n(t_i)+1}^{n} f_j(t_i)\right|\ell(I_i) \le \sum_{i\in d_2}\varepsilon\varphi(t_i)\ell(I_i) = \varepsilon S(\varphi, \mathcal{D}_2) \le \varepsilon,$$

where $\mathcal{D}_2 = \{(t_i, I_i) : i \in d_2\}$.

Next, from the non-negativity of the f_j,

$$T_2 \le \sum_{i=1}^{m}\sum_{j=n_0}^{n}\int_{I_i} f_j = \sum_{j=n_0}^{n}\int_{\mathbb{R}} f_j \le \sum_{j=n_0}^{\infty}\int_{\mathbb{R}} f_j < \varepsilon.$$

Finally, for T_1 set $s = \max\{n(t_i) : i \in d_2\}$ and note $\{(t_i, I_i) : n(t_i) = k\} <<$

γ_k. By Henstock's Lemma 3.1,

$$T_1 = \left| \sum_{i \in d_2} \left\{ \int_{I_i} s_{n(t_i)} - s_{n(t_i)}(t_i)\ell(I_i) \right\} \right|$$

$$= \left| \sum_{k=1}^{s} \sum_{\substack{i \\ n(t_i)=k}} \left\{ \int_{I_i} s_{n(t_i)} - s_{n(t_i)}(t_i)\ell(I_i) \right\} \right|$$

$$\leq \sum_{k=1}^{s} \left| \sum_{\substack{i \\ n(t_i)=k}} \left\{ \int_{I_i} s_{n(t_i)} - s_{n(t_i)}(t_i)\ell(I_i) \right\} \right| \leq \sum_{k=1}^{s} \varepsilon/2^k < \varepsilon .$$

From (1) it follows that $|\int_{\mathbb{R}} s_n - S(s_n, \mathcal{D})| < 4\varepsilon$ whenever $\mathcal{D} << \gamma$ and $n > n_0$. Hence, (i) is established.

Conditions (ii) and (iii) follow from Theorem 2.

From Theorem 4 we can immediately obtain one of the major convergence theorems for the gauge integral, the Monotone Convergence Theorem (sometimes called the Beppo Levi Theorem).

Theorem 5 (*Monotone Convergence Theorem*: *MCT*). *Let $f_k : I \to \mathbb{R}$ be integrable over $\mathbb{R}$ and suppose that $f_k(t) \uparrow f(t) \in \mathbb{R}$ for every $t \in \mathbb{R}$. If $\sup_k \int_I f_k < \infty$, then*

(i) *$\{f_k\}$ is uniformly integrable over I,*
(ii) *f is intcgrable over I and*
(iii) $\lim \int_I f_k = \int_I f (= \int_I (\lim f_k))$.

Proof: Set $f_0 = 0$ and $g_k = f_k - f_{k-1}$ for $k \geq 1$. Then $g_k \geq 0$, $\sum_{k=1}^{n} g_k = f_n \to f$ pointwise and $\sum_{k=1}^{\infty} \int_I g_k = \lim_n \sum_{k=1}^{n} \int_I (f_k - f_{k-1}) = \lim_n \int_I f_n = \sup_n \int_I f_n < \infty$. Hence, Theorem 4 is applicable and gives the result.

Of course, an analogous result holds for decreasing sequences (Exercise 3).

The Monotone Convergence Theorem gives a very useful and powerful sufficient condition guaranteeing "passage to the limit under the integral sign", i.e., $\lim \int f_k = \int (\lim f_k)$, but the monotone convergence requirement is often not satisfied. We next prove another convergence result, the Dominated Convergence Theorem, which relaxes this requirement. For this we require a preliminary result.

Definition 6. *A sequence $f_k : I \to \mathbb{R}$ is uniformly gauge Cauchy over I if for every $\varepsilon > 0$ there exist a gauge γ on I and an N such that $i, j \geq N$ implies $|S(f_i, \mathcal{D}) - S(f_j, \mathcal{D})| < \varepsilon$ whenever $\mathcal{D} << \gamma$.*

Proposition 7. *Let $f_k : I \to \mathbb{R}$ be integrable over I. Then $\{f_k\}$ is uniformly gauge Cauchy over I if and only if $\{f_k\}$ is uniformly integrable over I and $\lim \int_I f_k$ exists.*

Proof: Let $\varepsilon > 0$. Suppose $\{f_k\}$ is uniformly gauge Cauchy over I. Let γ be a gauge on I and N be such that $|S(f_i, \mathcal{D}) - S(f_j, \mathcal{D})| < \varepsilon/3$ whenever $i, j \geq N$ and $\mathcal{D} << \gamma$. Let $i, j \geq N$. Let γ_1 be a gauge on I such that

$$\left| \int_I f_i - S(f_i, \mathcal{D}) \right| < \varepsilon/3 \text{ and } \left| \int_I f_j - S(f_j, \mathcal{D}) \right| < \varepsilon/3$$

whenever $\mathcal{D} << \gamma_1$. Put $\gamma_2 = \gamma_1 \cap \gamma$. Then if $\mathcal{D} << \gamma_2$,

$$\left| \int_I f_i - \int_I f_j \right| \leq \left| \int_I f_i - S(f_i, \mathcal{D}) \right| + |S(f_i, \mathcal{D}) - S(f_j, \mathcal{D})| + \left| S(f_j, \mathcal{D}) - \int_I f_j \right| < \varepsilon$$

which implies $\{\int_I f_i\}$ is Cauchy and, therefore, convergent.

Pick $M \geq N$ such that $|\int_I f_i - \int_I f_j| < \varepsilon/3$ when $i, j \geq M$. Pick a gauge γ_3 on I such that

$$\left| S(f_M, \mathcal{D}) - \int_I f_M \right| < \varepsilon/3$$

whenever $\mathcal{D} << \gamma_3$. Put $\gamma_4 = \gamma_3 \cap \gamma_2$. If $i \geq M$ and $\mathcal{D} << \gamma_4$, then

$$\left| S(f_i, \mathcal{D}) - \int_I f_i \right| \leq |S(f_i, D) - S(f_M, \mathcal{D})| + \left| S(f_M, \mathcal{D}) - \int_I f_M \right| + \left| \int_I f_M - \int_I f_i \right| < \varepsilon$$

so $\{f_k\}$ is uniformly integrable.

For the converse, let γ' be a gauge on I such that $|S(f_i, \mathcal{D}) - \int_I f_i| < \varepsilon/3$ whenever $\mathcal{D} << \gamma'$ and $i \in \mathbb{N}$. Pick N such that $|\int_I f_i - \int_I f_j| < \varepsilon/3$ whenever

$i, j \geq N$. If $\mathcal{D} << \gamma'$ and $i, j \geq N$, then

$$|S(f_i, \mathcal{D}) - S(f_j, \mathcal{D})| \leq \left|S(f_i, \mathcal{D}) - \int_I f_i\right| + \left|\int_I f_i - \int_I f_j\right| + \left|\int_I f_j - S(f_j, \mathcal{D})\right| < \varepsilon$$

so $\{f_k\}$ is uniformly gauge Cauchy.

We now have the necessary machinery for the Dominated Convergence Theorem.

Theorem 8 (*Dominated Convergence Theorem: DCT*). *Let $f_k, f, g : I \to \mathbb{R}$ and assume f_k, g are integrable over I with $|f_i - f_j| \leq g$ for all i, j. If $\lim f_k = f$ pointwise on I, then*

(i) *$\{f_k\}$ is uniformly integrable over I,*
(ii) *f is integrable over I and*
(iii) $\lim \int_I f_k = \int_I f (= \int_I (\lim f_k))$.

Proof: Set $t_{j,k} = \vee\{|f_m - f_n| : j \leq m \leq n \leq k\}$. Then each $t_{j,k}$ is integrable by 4.11 and 4.12 and for each j, $\{t_{j,k}\}_{k=1}^{\infty}$ is an increasing sequence which converges to the function $t_j = \vee\{|f_m - f_n| : j \leq m \leq n\} \leq g$. The MCT implies that each t_j is integrable with $\int_I t_j \leq \int_I g$. Now $t_{j+1} \leq t_j$ and $t_j \to 0$ pointwise so the MCT implies $\int_I t_j \downarrow 0$.

Let $\varepsilon > 0$. There exists N such that $\int_I t_N < \varepsilon/2$ and there exists a gauge γ on I such that

$$\left|S(t_N, \mathcal{D}) - \int_I t_N\right| < \varepsilon/2$$

whenever $\mathcal{D} << \gamma$. If $i, j \geq N$ and $\mathcal{D} << \gamma$, then

$$|S(f_i, \mathcal{D}) - S(f_j, \mathcal{D})| \leq S(|f_i - f_j|, \mathcal{D}) \leq S(t_N, \mathcal{D}) \leq \int_I t_N + \varepsilon/2 < \varepsilon\,.$$

Hence, $\{f_i\}$ is uniformly gauge Cauchy on I and (i) follows from Proposition 7.

Conditions (ii) and (iii) follow from (i) and Theorem 2.

Remark 9. The "usual dominating hypothesis" in the DCT for the Lebesgue integral is that there exists an integrable function g such that $|f_j| \leq g$ for all j. Since $|f_i - f_j| \leq |f_i| + |f_j|$, this hypothesis clearly implies the one in Theorem 8 [see also Exercise 7]. On the other hand, the hypothesis $|f_i - f_j| \leq g$ allows the functions f_j to be conditionally integrable [just take any conditionally integrable function h and consider the sequence $\{f_j + h\}$] and is more appropriate for a conditional integral like the gauge integral. Exercise 5 gives another equivalent way of phrasing the dominating hypothesis in Theorem 8.

Besides the MCT and DCT another useful convergence theorem for bounded intervals is given in Exercise 1. [The analogue of the result in Exercise 1 for the Riemann integral is the "usual" convergence theorem for the Riemann integral.] See also Exercise 6.

We now give some examples illustrating the use of the convergence theorems. For this we derive some results pertaining to integrals which depend upon parameters. Let I be a closed subinterval of $\mathbb{R}^*$ and S a metric space (substitute $\mathbb{R}$ or $\mathbb{R}^n$ for S if necessary). If $f : S \times I \to \mathbb{R}$, we write $f(s,\cdot)[f(\cdot,t)]$ for the function $t \to f(s,t)[s \to f(s,t)]$ when $s \in S[t \in I]$. If $f(s,\cdot)$ is integrable over I for each $s \in S$, then $F(s) = \int_I f(s,t)dt$ defines a function $F : S \to \mathbb{R}$. Functions which depend upon parameters in this way arise in many areas of mathematics, and it is often important to ask if properties of the function $f(\cdot,t)$ are inherited by the function F. We first consider the property of continuity.

Theorem 10. *If $f(\cdot,t)$ is continuous at $s_0 \in S$ for each $t \in I$ and $f(s,\cdot)$ is integrable over I for each $s \in S$ and there is an integrable function $g : I \to \mathbb{R}$ such that $|f(s,t)| \le g(t)$ for all $t \in I$, then $F(s) = \int_I f(s,t)dt$ is continuous at s_0.*

Proof: Let $s_k \to s_0$ in S. Then $f(s_k,t) \to f(s_0,t)$ and $|f(s_k,t)| \le g(t)$ for every $t \in I$. The DCT implies that $F(s_k) \to F(s_0)$ so F is continuous at s_0.

Example 11. (*Gamma Function*) *The gamma function is defined by $\Gamma(x) = \int_0^\infty t^{x-1}e^{-t}dt$. We show that Γ is defined and continuous for $x > 0$. For $x > 0$ and $0 \le t \le 1, t^{x-1}e^{-t} \le t^{x-1}$ so Corollary 3.5 and Example 3.6 imply that the integral $\int_0^1 t^{x-1}e^{-t}dt$ exists. For $x > 0$ and $t > 1, t^{x-1}e^{-t} = [t^{x+1}e^{-t}]t^{-2}$. For $x > 0, \lim_{t\to\infty} t^{x+1}e^{-t} = 0$ so there is a constant M such that $t^{x-1}e^{-t} \le Mt^{-2}$. Now Corollary 4.5 and Example 4.6 imply that $\int_1^\infty t^{x-1}e^{-t}dt$ exists. Hence, $\Gamma(x)$ exists for $x > 0$.*

We next show that Γ is continuous. Fix $x_0 > 0$. If $0 \le t \le 1$ and $x_0/2 < x$, then $t^x \le t^{x_0/2}$ and $e^{-t} \le 1$ so $t^{x-1}e^{-t} \le t^{(x_0/2)-1}$. Hence, Theorem 10 implies that $\int_0^1 t^{x-1}e^{-t}dt$ is continuous at x_0. For $t \ge 1$ and $x_0/2 \le x \le 2x_0$, as above, there is a constant B such that $t^{x-1}e^{-t} \le Bt^{-2}$ so Theorem 10 again implies that $\int_1^\infty t^{x-1}e^{-t}dt$ is continuous at x_0.

For our next example we require a version of Leibniz' Rule for "differentiating under the integral sign."

Theorem 12 (*Leibniz' Rule*). *Let $S = [a, b]$. Suppose*

(i) $\frac{\partial f}{\partial s} = D_1 f$ *exist for all $s \in S$ and $t \in I$,*
(ii) *each $f(s, \cdot)$ is integrable over I and there is an integrable function $g : I \to \mathbb{R}$ such that*

$$\left|\frac{\partial f}{\partial s}(s,t)\right| = |D_1 f(s,t)| \le g(t)$$

for all $t \in I, s \in S$.

Then $F(s) = \int_I f(s,t)dt$ is differentiable on S with

$$F'(s) = \int_I \frac{\partial f}{\partial s}(s,t)dt = \int_I D_1 f(s,t)dt\,.$$

Proof: Fix $s_0 \in S$ and let $\{s_k\}$ be a sequence in S converging to s_0 with $s_k \neq s_0$. For $t \in I$

$$\lim_k (f(s_k,t) - f(s_0,t))/(s_k - s_0) = D_1 f(s_0,t) = \frac{\partial f}{\partial s}(s_0,t)$$

and the function

$$h_k(t) = \frac{f(s_k,t) - f(s_0,t)}{s_k - s_0}$$

is integrable over I. By the Mean Value Theorem, for each pair (k, t) there is a point $z_{k,t}$ lying between s_k and s_0 such that $h_k(t) = D_1 f(z_{k,t}, t)$ so $|h_k(t)| \le g(t)$. The DCT implies that $\lim_k h_k(t) = \int_I D_1 f(s_0,t)dt$, i.e., $F'(s_0) = \int_I D_1 f(s_0,t)dt$.

If I is a bounded interval and $D_1 f$ is continuous over $S \times I$, then $D_1 f$ is bounded and the function g in Theorem 12 can be taken to be a constant. We use this observation in the next example.

Example 13. *We show $A = \int_0^\infty e^{-t^2} dt = \sqrt{\pi}/2$. For this we introduce*

$$F(x) = \int_0^1 e^{-x(1+t^2)}/(1+t^2)dt\,.$$

Note that $F(0) =$ arctan $1 = \pi/4$, and if $x > 0$,

$$0 \le F(x) \le e^{-x} \int_0^1 \frac{1}{1+t^2} dt = e^{-x}\pi/4$$

so $F(\infty) = \lim_{x\to\infty} F(x) = 0$. *Fix* x. *We have, by Leibniz' Rule,*

$$F'(x) = -e^{-x}\int_0^1 e^{-xt^2}dt = -(e^{-x}/\sqrt{x})\int_0^{\sqrt{x}} e^{-u^2}du = -(e^{-x}/\sqrt{x})g(\sqrt{x}),$$

where $g(z) = \int_0^z e^{-u^2}du$. *Integrating* F' *from* 0 *to* ∞, *gives*

$$\begin{aligned}-\pi/4 &= -\int_0^\infty (e^{-x}/\sqrt{x})g(\sqrt{x})dx = -2\int_0^\infty e^{-z^2}g(z)dz \\ &= -2\int_0^\infty g'(z)g(z)dz = -g(\infty)^2 = -A^2\,.\end{aligned}$$

Hence, $A = \sqrt{\pi}/2$.

Notes/Remarks

Henstock originally established the MCT and DCT for the gauge integral by using what we have referred to as Henstock's Lemma (see the remarks in [H3] § 22). The formal definition of uniform integrability was introduced by Gordon ([Go2]), but it was used implicitly by McLeod in his proofs of the MCT and DCT ([ML] § 3.10). Schwabik used the notion under the name, equi-integrability and also established Theorem 2 ([Sch]). If $\{f_k\}$ is a pointwise convergent sequence of gauge integrable functions, Gordon gives a characterization of uniform integrability of $\{f_k\}$ in terms of their indefinite integrals ([Go] 13.29). It will be shown in Theorem 7.8 that the convergence conclusion of Theorem 3, the MCT and the DCT can be improved to converge in the Alexiewicz norm.

Bartle has given an interesting condition (called γ-convergence) for the limit f of a sequence of gauge integrable functions $\{f_k\}$ to be integrable and for $\lim \int f_k = \int f$ to hold ([Ba2]). See also [Go3] where Gordon discusses γ-convergence and alters Bartle's definition slightly.

There is a more general convergence theorem for the gauge integral called the Controlled Convergence Theorem; this result implies both the MCT and DCT. However, this theorem involves more technical notions than we are assuming in this text. The interested reader can consult [LPY], Chapter 7, for the statement and several proofs of the Controlled Convergence Theorem; there are also a number of references given in the bibliography.

The derivation in Example 13 is from [We]; a generalization of the formula in Example 13 is given in [DS], Example 15.10.

Exercise 33 gives an interesting improvement to Theorem 3.4 for "improper integrals."

Exercises

1. Suppose I is a bounded interval and $f_k : I \to \mathbb{R}$ is integrable over I with $f_k \to f$ uniformly on I. Show that

   (i) $\{f_k\}$ is uniformly integrable over I,
   (ii) f is integrable over I and
   (iii) $\int_I f_k \to \int_I f$.

2. Show that the analogue of the MCT (DCT) is false for the Riemann integral.
3. State and prove a version of the MCT for decreasing sequences $\{f_k\}$.
4. Let $f_k, f : I \to \mathbb{R}$ be non-negative with each f_k integrable over I and $f = \sum_{k=1}^{\infty} f_k$ pointwise.
   Give necessary and sufficient conditions for f to be integrable over I.
5. Let $f_k : I \to \mathbb{R}$ be integrable over I. Show that there exists an integrable function $g : I \to \mathbb{R}$ satisfying $|f_k - f_j| \leq g$ for all k, j if and only if there exist integrable functions g and h satisfying $g \leq f_k \leq h$ for all k.
6. (Bounded Convergence Theorem: BCT) Let I be a closed, bounded interval and let f_k be integrable over I with $f_k \to f$ pointwise. If there exists $M \geq 0$ such that $|f_k(t)| \leq M$ for all k, t, show that

   (i) $\{f_k\}$ is uniformly integrable over I,
   (ii) f is integrable over I and
   (iii) $\int_I f_k \to \int_I f$.

7. Let $f_k : I \to \mathbb{R}$ be integrable over I and $f_k \to f$ pointwise. Suppose there exists an integrable function g such that $|f_k| \leq g$ for all k. Show that in this case the conclusion of the DCT can be improved to read $\int_I |f_k - f| \to 0$.
8. Let f_k, f, g satisfy the hypothesis in Exercise 7 and let $I = [a, b]$. If $F_k(t) = \int_a^t f_k$, $F(t) = \int_a^t f$, show that $F_k \to F$ uniformly on I.
9. (Fatou's Lemma) Let $f_k : I \to \mathbb{R}$ be non-negative and integrable and assume that $\underline{\lim} f_k$ is finite on I. Show that $\underline{\lim} \int_I f_k \leq \int_I (\underline{\lim} f_k)$. Give examples showing strict inequality may hold and the non-negativity assumption cannot be dropped. [Hint: Consider $h_k = \inf\{f_j : j \geq k\}$ and use MCT.]
10. Give an example for which $f_k \to 0$ pointwise, $\int_I f_k \to 0$ but $\{f_k\}$ is not dominated by an integrable function.
11. Show that $F(x) = \int_1^{\infty} \frac{\sin t}{x^2 + t^2} dt$ is continuous for $x \in \mathbb{R}$.
12. Show that $F(x) = \int_0^{\infty} e^{-xt} \sin t \; dt$ defines a continuous function for $x > 0$.
13. Show that

    (i) $\Gamma(x + 1) = x\Gamma(x)$ for $x > 0$,
    (ii) $\Gamma(n + 1) = n!$ for $n \in \mathbb{N}$,

(iii) $\lim_{x\to 0^+} x\Gamma(x) = 1$,
(iv) $\Gamma(\frac{1}{2}) = \sqrt{\pi}$.
Show that Γ has derivatives of all orders and give a formula for $\Gamma^{(k)}$.

14. Show that $\int_0^\infty x^{2n}e^{-x^2}dx = \frac{(2n)!\sqrt{\pi}}{2^{2n}n!2}$ for $n = 0, 1, 2, \ldots$. [Hint: $n = 0$ is Example 13.]
15. Show that $\int_0^\infty e^{-tx^2}dx = \frac{1}{2}\sqrt{\pi/t}$ for $t > 0$.
16. Show that $F(x) = \int_0^\infty \frac{e^{-xt}}{1+t}dt$ is differentiable for $x > 0$.
17. Show that $F(x) = \int_0^\infty \frac{e^{-t}}{1+xt}dt$ is differentiable for $0 \le x \le 1$.
18. Let $f : [0,1] \to \mathbb{R}$ be continuous. Show that $\int_0^1 f(t^k)dt \to f(0)$.
19. Let $f : [0,\infty) \to \mathbb{R}$ be continuous and $\lim_{t\to\infty} f(t) = L$. What can you say about $\lim_k \int_0^2 f(kt)dt$?
20. Evaluate $\varphi(x) = \int_0^1 (t^x - 1)/\ln t\, dt$ for $x > 0$. [Hint: Find $\varphi'(x)$ and note $\varphi(x) \to 0$ as $x \to 0^+$.]
21. Let $f_k(t) = (1/k^2\sqrt{t})\cos(k/t)$. Show that $\sum_{k=1}^\infty \int_0^1 f_k = \int_0^1 \sum_{k=1}^\infty f_k$.
22. Let $f_k(t) = e^{-kt} - 2e^{-2kt}$. Show that $\sum_{k=1}^\infty \int_0^\infty f_k \neq \int_0^\infty \sum_{k=1}^\infty f_k$.
23. Let $f : [-1,1] \to \mathbb{R}$ be continuous. Show that $\lim_{x\to 0^+} \int_{-1}^1 \frac{x}{x^2+t^2} f(t)dt = \frac{\pi}{2}f(0)$.
24. Let $f : [0,\infty) \to \mathbb{R}$ be bounded and continuous. Show that $\lim_k \int_0^\infty \frac{f(t)}{1+kt^p}dt = 0$ for $p > 1$.
25. Let $f : [a,b] \times [c,d] \to \mathbb{R}$ be continuous. Show that $F(x) = \int_c^d f(x,y)dy$ is continuous for $a \le x \le b$.
26. Show that $\lim_{x\to 0^+} \int_0^1 \frac{e^{-xt}}{1+t}dt = \ln 2$.
27. Evaluate $\lim_{x\to\infty} \int_0^\infty \frac{e^{-xt}}{1+t}dt$.
28. Let $f : \mathbb{R} \to \mathbb{R}$ be absolutely integrable and uniformly continuous on $\mathbb{R}$. Show that $\lim_{|x|\to\infty} f(x) = 0$. Can "uniform continuity" be replaced by "continuity"?
29. (Riemann-Lebesgue Lemma: junior grade) Let $f : [0,1] \to \mathbb{R}$ be continuous. Show that $\lim_k \int_0^1 f(t)\sin(kt)dt = 0$. [Hint: First consider the case when f is a step function.]
30. (Laplace Transform) Let $f : [0,\infty] \to \mathbb{R}$ be integrable over every bounded subinterval. The *Laplace transform* of f is defined by $\mathcal{L}\{f\}(s) = \int_0^\infty e^{-st} f(t)dt$. The function f is of *exponential order* if there exist constants a, M such that $|f(t)| \le Me^{at}$ for $t > 0$. Show that the Laplace transform of such a function exists and defines a continuous function for $s > a$.
31. Evaluate $\lim_k \int_0^k (1 + t/k)^k e^{-2t}dt$.
32. Suppose $f : [a,b] \to \mathbb{R}$ is continuous and $\int_a^b t^k f(t)dt = 0$ for $k = 0, 1, 2, \ldots$. Show that $f = 0$. [Hint: Weierstrass Approximation Theorem.]

33. (See Theorem 3.4) Let $f : [a,b] \to \mathbb{R}$ be gauge integrable over $[a,b]$ and let $a_k \downarrow a, a < a_k < b$. Define f_k by $f_k(t) = f(t)$ for $a_k \le t \le b$ and $f_k(t) = 0$ for $a \le t < a_k$. Show that $\{f_k\}$ is uniformly integrable on $[a,b]$. Hint: Set $F(t) = \int_a^t f$. Pick $\delta > 0$ such that $|F(t) - F(s)| < \varepsilon$ for $|t - s| < \delta$. Pick γ such that $\gamma(t) \subset (t - \delta, t + \delta)$ and $|S(f, \mathcal{D}) - \int_a^b f| < \varepsilon$ whenever $\mathcal{D} << \gamma$. Suppose $\mathcal{D} << \gamma$. For each k let $\mathcal{D}_k$ be the subset of $\mathcal{D}$ with tags greater/equal a_k. The union of the subintervals in $\mathcal{D}_k$ is an interval $[y,b]$ with $|y - a_k| < \delta$. Use the Henstock Lemma 3.1 to show

$$\left| S(f_k, \mathcal{D}) - \int_a^b f_k \right| < 2\varepsilon .$$

34. Use Exercise 33 to give an example of a sequence which is uniformly integrable but does not satisfy the hypothesis in either MCT or DCT. [Hint: Use a conditionally convergent integral.]
35. Let $f_k(0) = k$ and $f_k(t) = 0$ for $0 < t \le 1$. Show that $\{f_k\}$ is not uniformly integrable over $[0,1]$.
36. Let $f, g : I \to [0, \infty)$. Assume that f is integrable over I and g is integrable over every bounded subinterval of I. Show that $f \wedge g$ is integrable over I. In particular, $f \wedge k$ is integrable over I for every $k \in \mathbb{N}$. [Hint: Use MCT.]
37. Let $f : I \to \mathbb{R}$ be absolutely integrable over I. For $k \in \mathbb{N}$ set $f^k(t) = k$ if $f(t) < -k$, $f^k(t) = k$ if $f(t) > k$ and $f^k(t) = f(t)$ if $|f(t)| \le k$ [the truncation of f at k]. Show that f^k is absolutely integrable over I. [Hint: consider $f \wedge k = g$ and $h = (-k) \vee g$.]

Chapter 6

Integration over More General Sets: Lebesgue Measure

In previous chapters we considered integrating functions which are gauge integrable over subintervals (or finite unions of subintervals). In this chapter we consider integrating functions which are gauge integrable over more general subsets. The sets which we consider are the Lebesgue measurable subsets. This chapter uses only results from previous chapters but is not necessary for reading the unstarred chapters which follow.

There is a natural notion of length for subintervals of $\mathbb{R}$, and we now seek to extend the length function in a natural way to certain subsets of $\mathbb{R}$. We say that a subset E of $\mathbb{R}$ is (Lebesgue) *integrable* if C_E is (gauge) integrable over $\mathbb{R}$ and define the (Lebesgue) *measure* of E to be $\lambda(E) = \int_{\mathbb{R}} C_E$. Let $\mathcal{L}$ be the family of all integrable subsets of $\mathbb{R}$. We have the following properties of $\mathcal{L}$ and the set function λ.

Proposition 1. (*i*) *All bounded intervals I belong to $\mathcal{L}$ and $\lambda(I) = \ell(I)$. In particular, $\phi \in \mathcal{L}$ and $\lambda(\phi) = 0$.*

(ii) *If $E, F \in \mathcal{L}$, then $E \cup F$, $E \cap F$ and $E \backslash F$ belong to $\mathcal{L}$.*

(iii) *If $E_k \in \mathcal{L}$ for every $k \in \mathbb{N}$, then $\bigcap_{k=1}^{\infty} E_k \in \mathcal{L}$.*

(iv) *$E \in \mathcal{L}$ if and only if $E + a \in \mathcal{L}$ for every $a \in \mathbb{R}$ and, in this case $\lambda(E) = \lambda(E + a)$ (translation invariance).*

(v) *If $\{E_i\} \subset \mathcal{L}$ is pairwise disjoint and $E = \bigcup_{i=1}^{\infty} E_i$, then $E \in \mathcal{L}$ if and only if $\sum_{i=1}^{\infty} \lambda(E_i) < \infty$ and, in this case, $\lambda(E) = \sum_{i=1}^{\infty} \lambda(E_i)$ (countable additivity).*

(vi) *If $E, F \in \mathcal{L}$ and $E \subset F$, then $\lambda(E) \le \lambda(F)$.*

(vii) *If E is null, then $E \in \mathcal{L}$ and $\lambda(E) = 0$.*

Proof: (i) is clear. (ii) follows from the identities $C_{E\cup F} = C_E \vee C_F, C_{E\cap F} = C_E \wedge C_F$ and $C_{E\setminus F} = C_E - C_{E\cap F}$ (Proposition 4.12).

For (iii), let $E = \bigcap_{k=1}^{\infty} E_k$ and $F_k = \bigcap_{i=1}^{k} E_i$. Then $F_k \in \mathcal{L}$ by (ii) and $C_{F_k} \downarrow C_E$ so C_E is integrable by MCT (Theorem 5.5), i.e., $E \in \mathcal{L}$.

(iv) follows from Exercise 2.12.

For (v), $C_E = \sum_{k=1}^{\infty} C_{E_k}$ so this part is immediate from MCT.

(vi) follows from (ii) and (v).

(vii) follows from Example 1.10 or Exercise 1.9.

We say that a subset $E \subset \mathbb{R}$ is (Lebesgue) *measurable* if $E \cap I$ is integrable for every bounded interval I. We denote the family of all measurable subsets of $\mathbb{R}$ by $\mathcal{M}$ and note that $\mathcal{L} \subset \mathcal{M}$ by Proposition 1. We extend λ to $\mathcal{M}$ by setting $\lambda(E) = \infty$ if $E \in \mathcal{M} \setminus \mathcal{L}$ and call λ *Lebesgue measure* on $\mathbb{R}$.

We have the following properties for $\mathcal{M}$ and λ.

Proposition 2. *(i)* $\phi \in \mathcal{M}$ *and* $\lambda(\phi) = 0$.

(ii) $E \in \mathcal{M}$ *implies* $E^c = \mathbb{R} \setminus E \in \mathcal{M}$.

(iii) $E_i \in \mathcal{M}$ *for* $i \in \mathbb{N}$ *implies* $\bigcup_{i=1}^{\infty} E_i, \bigcap_{i=1}^{\infty} E_i \in \mathcal{M}$.

(iv) *If* I *is an interval (bounded or unbounded),* $\lambda(I) = \ell(I)$.

(v) *If* $\{E_i\} \subset \mathcal{M}$ *is pairwise disjoint, then* $\lambda(\bigcup_{i=1}^{\infty} E_i) = \sum_{i=1}^{\infty} \lambda(E_i)$ *(countable additivity).*

(vi) $E \in \mathcal{M}$ *if and only if* $E + a \in M$ *for every* $a \in \mathbb{R}$ *and, in this case,* $\lambda(E) = \lambda(E + a)$ *(translation invariance).*

Proof: (i) is clear. For (ii) let I be a bounded interval. Then $I = (I \cap E) \cup (I \cap E^c)$ and $I \in \mathcal{L}$, $I \cap E \in \mathcal{L}$ imply $I \cap E^c \in \mathcal{L}$ by Proposition 1 (ii).

For (iii), if I is a bounded interval and $A, B \in \mathcal{M}$, then $I \cap (A \cup B) = (I \cap A) \cup (I \cap B)$ and $I \cap (A \cap B) = (I \cap A) \cap (I \cap B)$ so $A \cup B$ and $A \cap B$ belong to $\mathcal{M}$ by Proposition 1. If $\{E_i\} \subset \mathcal{M}$, we construct a pairwise disjoint sequence $\{F_k\}$ from $\mathcal{M}$ whose union is $E = \bigcup_{i=1}^{\infty} E_i$. For this, set $F_1 = E_1$ and $F_{k+1} = E_{k+1} \setminus \bigcup_{i=1}^{k} E_i$. Then the $\{F_k\}$ are pairwise disjoint, $F_k \in \mathcal{M}$ by (ii) and the observation above and $E = \bigcup_{k=1}^{\infty} F_k$. From Proposition 1 $\sum_{k=1}^{n} \lambda(F_k \cap I) \le \lambda(I)$ for each n so $\sum_{k=1}^{\infty} \lambda(F_k \cap I) \le \lambda(I)$ and $E \cap I \in \mathcal{L}$, i.e., $E \in \mathcal{M}$. Since $\bigcap_{i=1}^{\infty} E_i = (\bigcup_{i=1}^{\infty} E_i^c)^c = F$, $F \in \mathcal{M}$ by (ii) and what has just been established.

(iv) is clear, (v) follows from Proposition 1 (v) and (vi) follows from Proposition 1 (iv).

At this point it is not clear how large the families $\mathcal{L}$ and $\mathcal{M}$ are. Since every open subset of $\mathbb{R}$ is a countable union of pairwise disjoint open intervals ([DS] 16.6), it follows from Proposition 2 that every open set is measurable and its measure is the sum of the lengths of the open intervals whose union is the open set. Also, from Exercise 3 every bounded open set is integrable. It now follows from Proposition 2 (ii) that every closed subset of $\mathbb{R}$ is also measurable.

We now introduce some useful terminology from measure theory. A family of subsets $\sum$ of a set S is called a *σ-algebra* if $\sum$ satisfies conditions (i)–(iii) of Proposition 2. If $\mathcal{F}$ is any family of subsets of a set S, there is a smallest σ-algebra containing $\mathcal{F}$ called the σ-algebra generated by $\mathcal{F}$ (Exercise 1). The σ-algebra $\mathcal{B} = \mathcal{B}(\mathbb{R})$ generated by the open subsets of $\mathbb{R}$ is called the family of *Borel sets* of $\mathbb{R}$. From Proposition 2 and the observations above, we see that $\mathcal{M}$ is a σ-algebra containing the Borel sets $\mathcal{B}$. It is the case that $\mathcal{B} \neq \mathcal{M}$ and assuming the Axiom of Choice that $\mathcal{M} \neq \mathcal{P}(\mathbb{R})$, where $\mathcal{P}(\mathbb{R})$ is the power set of $\mathbb{R}$ (see Notes/Remarks for references to these results). We will show later (Theorem 5 and Corollaries 7,8) that every measurable set is approximated in some sense by a Borel set.

A set function ν defined on a σ-algebra $\sum$ is said to be *countably additive* if $\nu : \sum \to \mathbb{R}^*$ takes on at most one of the values $\{\pm\infty\}$, $\nu(\phi) = 0$ and $\nu(E) = \sum_{k=1}^{\infty} \nu(E_k)$ [convergence in $\mathbb{R}^*$], whenever $\{E_k\}$ is a pairwise disjoint sequence from $\sum$ with $E = \bigcup_{k=1}^{\infty} E_k$. A *countably additive* set function ν is said to be a *measure* if ν is non-negative. It follows from Proposition 2 that λ is a measure — called *Lebesgue measure* — on $\mathcal{M}$ which is translation invariant in the sense of Proposition 2 (vi). [It is the case that Lebesgue measure is essentially the only such measure which extends the length function; see Notes/Remarks for references.]

We now proceed to give some more geometric descriptions of the class of measurable sets. For this we first establish a useful covering lemma.

Lemma 3 (*Covering Lemma*). *Let I be a closed, bounded interval in $\mathbb{R}, \phi \neq E \subset I$ and γ a gauge on I. There exists a family $\{(t_k, J_k) : k \in M\}$ which is at most countable where each J_k is a closed subinterval of I and $J_k^0 \cap J_\ell^0 = \emptyset$ if $k \neq \ell$, $t_k \in J_k \cap E$, $J_k \subset \gamma(t_k)$ and $E \subset \bigcup_{k \in M} J_k \subset I$.*

Proof: Let $\mathcal{D}_k$ be the closed subintervals of I obtained by dividing I into 2^k equal subintervals (i.e., bisect I and keep bisecting). Let $\mathcal{E}_1$ be the family of $J \in \mathcal{D}_1$ such that there exists $t \in E \cap J$ with $J \subset \gamma(t)$. Let $\mathcal{E}_2$ be the family of all $J \in \mathcal{D}_2$ such that J is not contained in an element of $\mathcal{E}_1$ and there exists $t \in E \cap J$ with $J \subset \gamma(t)$. Continuing this procedure produces a sequence $(\mathcal{E}_k)$ of families of closed subintervals of I (some $\mathcal{E}_k$ may be empty). Then $\mathcal{E} = \bigcup_{k=1}^{\infty} \mathcal{E}_k$ is an at most countable family of non-overlapping, closed intervals of I. We claim that $E \subset \cup\{J : J \in \mathcal{E}\}$. Let $t \in E$. There exists an integer k_1 such that if $k \geq k_1$, then the closed interval $J_k(t)$ of $\mathcal{D}_k$ containing

t must be contained in $\gamma(t)$. Then either $J_{k_1}(t) \in \mathcal{E}_{k_1}$ or $J_{k_1}(t) \in \bigcup_{k=1}^{k_1-1} \mathcal{E}_k$. Hence, $E \subset \cup\{J : J \in \mathcal{E}\}$. Arrange the elements of $\{J_k(t) : t \in E\}$ into a (possibly finite) sequence $\{J_k : k \in M\}$. Then for each $k \in M$ there exists $t_k \in E \cap J_k$ such that $J_k \subset \gamma(t_k)$. Then $\{(t_k, J_k) : k \in M\}$ is the desired family.

We can now establish an approximation theorem for bounded measurable sets.

Theorem 4. *Let E be a bounded, measurable set contained in the bounded, closed interval I and let $\varepsilon > 0$. There exists an at most countable family $\{J_k : k \in M\}$ of closed, non-overlapping subintervals of I such that $E \subset \cup\{J_k : k \in M\}$ and $\sum_{k \in M} \ell(J_k) \leq \lambda(E) + \varepsilon$.*

Proof: There exists a gauge γ on I such that $|S(C_E, \mathcal{D}) - \lambda(E)| < \varepsilon$ whenever $\mathcal{D} << \gamma$ (Exercise 3). Let $\{(t_k, J_k) : k \in M\}$ be the family in Lemma 3 relative to ε, E and the gauge γ, where we assume $M \subset \mathbb{N}$. By Henstock's Lemma for every $q \in M$,

$$\left|\sum_{k=1}^{q}\{\lambda(E \cap J_k) - C_E(t_k)\ell(J_k)\}\right| = \left|\sum_{k=1}^{q}\{\lambda(E \cap J_k) - \ell(J_k)\}\right| \leq \varepsilon \,.$$

Therefore, for every $q \in M$, $\sum_{k=1}^{q} \ell(J_k) \leq \sum_{k=1}^{q} \lambda(E \cap J_k) + \varepsilon \leq \lambda(E) + \varepsilon$, and, hence, $\sum_{k \in M} \ell(J_k) \leq \lambda(E) + \varepsilon$ so the result follows.

From Theorem 4 it is now easy to obtain an approximation theorem for general measurable subsets.

Theorem 5. *Let E be measurable. For every $\varepsilon > 0$ there exists an open set $G \supset E$ such that $\lambda(G \backslash E) < \varepsilon$.*

Proof: First assume that E is bounded and is contained in a closed, bounded interval I. We can extend each interval J_k in Theorem 4 to be an open interval, still denoted by J_k, to be such that $\sum_{k \in M} \ell(J_k) \leq \lambda(E) + \varepsilon$ [replace ε by $\varepsilon/2$ and extend each J_k by a factor of $\varepsilon/2^{k+1}$]. Set $G = \cup\{J_k : k \in M\}$ so G is open and $E \subset G$. Also, by Exercise 2,

$$\lambda(E) \leq \lambda(G) \leq \sum_{k \in M} \lambda(J_k) \leq \lambda(E) + \varepsilon \,,$$

and by Exercise 2,

$$\lambda(G \setminus E) = \lambda(G) - \lambda(E) \leq \varepsilon \,.$$

If E is measurable, let $E_k = E \cap [-k, k]$. By the first part, for every k there exists an open set $G_k \supset E_k$ such that $\lambda(G_k \setminus E_k) < \varepsilon/2^k$. Set $G = \bigcup_{k=1}^{\infty} G_k$ so G is an open set containing $E = \bigcup_{k=1}^{\infty} E_k$. Since $G \setminus E \subset \bigcup_{k=1}^{\infty}(G_k \setminus E_k)$, Exercise 2 gives

$$\lambda(G \setminus E) \le \sum_{k=1}^{\infty} \lambda(G_k \setminus E_k) < \sum_{k=1}^{\infty} \varepsilon/2^k = \varepsilon\,.$$

We now derive several corollaries from Theorem 5. The first gives a useful characterization of null sets.

Corollary 6. *E is null if and only if E is integrable and $\lambda(E) = 0$.*

Proof: $\Rightarrow$: follows from Example 1.10 or Exercise 1.9.

$\Leftarrow$: Let $\varepsilon > 0$. By the argument in Theorem 5 for every k there exists a sequence of open intervals $\{J_j^k : j \in \mathbb{N}\}$ such that $E_k \subset \bigcup_{j=1}^{\infty} J_j^k$ and $\sum_{j=1}^{\infty} \ell(J_j^k) < \varepsilon/2^k$. Then $\{J_j^k : k, j \in \mathbb{N}\}$ covers E and $\sum_{k=1}^{\infty} \sum_{j=1}^{\infty} \ell(J_j^k) < \varepsilon$ so E is null.

A subset $H \subset \mathbb{R}$ is called a $\mathcal{G}_\delta$ if H is a countable intersection of open sets.

Corollary 7. *Let E be measurable. Then there exists a $\mathcal{G}_\delta$ set $H \supset E$ such that $\lambda(H \setminus E) = 0$.*

Proof: By Theorem 5 for every $k \in \mathbb{N}$ there exists an open set $G_k \supset E$ such that $\lambda(G_k \setminus E) < 1/k$. Put $H = \bigcap_{k=1}^{\infty} G_k$. Then H is a $\mathcal{G}_\delta$, and since $H \setminus E \subset G_k \setminus E$ for every k, $\lambda(H \setminus E) < 1/k$ so $\lambda(H \setminus E) = 0$.

Corollary 8. *$E \subset \mathbb{R}$ is measurable if and only if $E = B \cup Z$, where B is a Borel set and Z is null.*

Proof: Since $\mathcal{B} \subset \mathcal{M}$ and any null set is measurable, the condition is sufficient.

Conversely, if E is measurable, let H be as in Corollary 7 and let $Z = H \setminus E$. Since $H \in \mathcal{B}$, the result follows from Corollary 7.

We now establish a result which illustrates the utility of the class of measurable sets. We say that a function $f : I \to \mathbb{R}$ is integrable over a subset $E \subset I$ if fC_E is integrable over I and set

$$\int_E f = \int_I fC_E\,.$$

Theorem 9. *Let I be a closed interval in $\mathbb{R}^*$ and $f : I \to \mathbb{R}$. If f is absolutely integrable over I, then f is (absolutely) integrable over every measurable subset* *of I. Moreover, if $\mathcal{M}_I$ is the σ-algebra of measurable subsets of I and $\nu(E) = \int_E f$ for $E \in \mathcal{M}_I$, then ν is countably additive.*

Proof: Since $f = f^+ - f^-$ and both f^+ and f^- are integrable over I, we may assume that f is non-negative. Let $E \in \mathcal{M}_I$. Now $f_k = f \wedge (kC_{[-k,k]\cap E})$ is integrable over I for every k (Proposition 4.12), $f_k \uparrow fC_E$ and $0 \le f_k \le f$ so fC_E is integrable by DCT (Theorem 5.8).

Suppose $\{E_k\}$ is a pairwise disjoint sequence from $\mathcal{M}_I$ with $E = \bigcup_{k=1}^{\infty} E_k$. Then $fC_E = \sum_{k=1}^{\infty} fC_{E_k}$ so $\nu(E) = \sum_{k=1}^{\infty} \nu(E_k)$ by the MCT (Theorem 5.5).

The absolute integrability assumption in Theorem 9 is important; see Exercise 8. See also Theorem 13 in Notes/Remarks.

See Exercise 12 for a partial converse to Theorem 9.

We now give improvements in the MCT and DCT which are sometimes useful in applications. For the MCT we need a preliminary lemma.

Lemma 10. *Let $f_k : I \to \mathbb{R}$ be integrable over I for every k with $0 \le f_k \le f_{k+1}$. If $M = \sup \int_I f_k < \infty$ and if $f(t) = \lim f_k(t)$ [limit in $\mathbb{R}^*$], then f is finite-valued a.e. in I.*

Proof: Let $E = \{t \in I : f(t) = \infty\}$. For each j and k, f_k/j is integrable so $(f_k/j) \wedge 1$ is integrable [Exercise 5.36] with

$$\int_I (f_k/j) \wedge 1 \le \int_I (f_k/j) \le M/j\,.$$

For each

$$j(f_k/j) \wedge 1 \uparrow (f/j) \wedge 1$$

as $k \to \infty$ so the MCT implies

$$(*) \lim_k \int_I (f_k/j) \wedge 1 = \int_I (f/j) \wedge 1 \le M/j\,.$$

Now $f \ge 0$ so $(f/j) \wedge 1 \downarrow$ and $\lim_j \int_I (f/j) \wedge 1 = 0$ by (*). If $t \in I \setminus E$, $(f(t)/j) \wedge 1 = f(t)/j$ for large j so $(f(t)/j) \wedge 1 \to 0$; if $t \in E$, $(f(t)/j) \wedge 1 = 1$ so $(f(t)/j) \wedge 1 \to 1$. Therefore, $(f/j) \wedge 1 \to C_E$ and by the MCT, C_E is integrable with $\int_I C_E = 0$. Corollary 6.6 now gives the conclusion.

We can now establish a generalization of the MCT.

Theorem 11 (*General MCT*). *Let $f_k : I \to \mathbb{R}$ be integrable over I with $f_k \le f_{k+1}$ and $\sup_k \int_I f_k < \infty$. Then there exists an integrable function $f : I \to \mathbb{R}$ such that $f_k \to f$ a.e. in I and $\int_I f_k \uparrow \int_I f$.*

Proof: By replacing f_k by $f_k - f_1$, if necessary, we may assume $f_k \geq 0$. Let $E = \{t \in I : \lim f_k(t) < \infty\}$. Put $g_k = C_E f_k$ so $f_k = g_k$ a.e. and $0 \leq g_k \leq g_{k+1}$ by Lemma 10.

Let $f = \lim g_k$ and note $f(t) < \infty$ for every $t \in I$. The MCT and Exercise 2.22 implies that $\lim \int_I g_k = \lim \int_I f_k = \int_I f$.

We can also give a more general form of the DCT.

Theorem 12 (*General DCT*). *Let $f_k, f, g : I \to \mathbb{R}$. Assume that $f_k, k \in \mathbb{N}$ and g are integrable over I with $f_k \to f$ a.e. and $|f_k| \leq g$ a.e. Then f is integrable over I with* $\lim \int_I f_k = \int_I f$.

We leave the proof for Exercise 13.

Notes/Remarks

The construction of Lebesgue measure can be legitimately viewed as an attempt to extend the natural length function defined on subintervals of $\mathbb{R}$ to a more general class of subsets of $\mathbb{R}$. Of course, such an extension should possess some of the properties of the natural length function; for example, it is natural to require that such an extension satisfy the translation invariance property of Proposition 2 and also the countable additivity property of Proposition 2. We have obtained such an extension by using our construction of the gauge integral, but it would be much more desirable to have a purely geometric construction of the extension. There are several such constructions which are "well-known"; see for example, [Sw1] 2.5, [N1] III, [Ro] § 3. In these extensions, Theorem 5, Corollary 7 and their converses hold and give geometric characterizations of the class of (Lebesgue) measurable sets (the class of subsets of $\mathbb{R}$ to which the length function is extended). It is known that such an extension is essentially unique ([Sw1] 2.5.6), and if $\mathcal{M}$ is again the class of measurable subsets, then $\mathcal{B} \subsetneq \mathcal{M} \subsetneq \mathcal{P}(\mathbb{R})$ ([Sw1] 2.5.9, 1.3.1). See also [N1], [Sw1] 2.5 for other interesting properties of Lebesgue measure.

Exercise 8 shows that the absolute integrability of f in Theorem 9 is important. In fact, the converse of Theorem 9 holds; however, its proof seems to require the measurability of a gauge integrable function.

Theorem 13*. *Let I be a closed interval in $\mathbb{R}^*$ and $f : I \to \mathbb{R}$. If f is integrable over every measurable subset of I, then f is absolutely integrable over I.*

Proof: Since f is measurable (Appendix 3),

$$P = \{t \in I : f(t) \geq 0\} \text{ and } N = \{t \in I : f(t) < 0\}$$

are measurable. Therefore, $f^+ = fC_P$ and $f^- = fC_N$ are integrable over I by hypothesis and $|f| = f^+ + f^-$ is integrable over I.

It follows from Theorem 13 that a function which is integrable over an interval I is not, in general, integrable over every measurable subset of I. In fact, a function which is integrable over an interval I needn't be integrable over a countable disjoint union of subintervals of I ([LPY] p. 115). Theorem 3 of [YL] gives a necessary and sufficient condition for a function which is integrable over an interval I to be integrable over a pairwise disjoint union of subintervals of I. A more general result guaranteeing the integrability over a pairwise disjoint union of measurable subsets is given in Theorem 7 of [LPY1].

Exercises

1. If $\mathcal{F}$ is any family of subsets of a set S, show that there is a smallest σ-algebra containing $\mathcal{F}$.
2. Let $\sum$ be a σ-algebra of subsets of a set S and assume $\mu : \sum \to [0, \infty]$ is a measure. Show: (i) if $E, F \in \sum, E \subset F$, then $\mu(E) \leq \mu(F)$ [monotonicity], and if $\mu(E) < \infty$, then $\mu(F \backslash E) = \mu(F) - \mu(E)$, and (ii) if $\{E_i\} \subset \sum$, then $\mu(\bigcup_{i=1}^{\infty} E_i) \leq \sum_{i=1}^{\infty} \mu(E_i)$ [countable subadditivity]. [Hint: For (ii), disjointify the $\{E_i\}$ as in the proof of Proposition 2 (iii).]
3. Show that every bounded measurable set in $\mathbb{R}$ is integrable.
4. Let $f : \mathbb{R} \to \mathbb{R}$ and A, B be measurable with $A \subset B$. Show that if f is absolutely integrable over B, then f is absolutely integrable over A. Show the absolute integrability hypothesis cannot be replaced by integrability. [Hint: For $f \geq 0$, consider $f_k = f \wedge kC_{A \cap [-k,k]}$; Exercise 3.9 or Exercise 4.3.]
5. Give an example of a Borel set which is neither open or closed.
6. Let E be measurable and $\varepsilon > 0$. Show that there exists a closed (compact) subset $F \subset E$ such that $\lambda(E \setminus F) < \varepsilon$.
7. A subset H of $\mathbb{R}$ is called an $\mathcal{F}_\sigma$ set if H is a countable union of closed sets. If E is measurable, show that there exists an $\mathcal{F}_\sigma$ set $H \subset E$ such that $\lambda(E \setminus H) = 0$.
8. Show that the absolute integrability assumption in Theorem 9 cannot be replaced by integrability. [Hint: Exercise 3.9 or Exercise 4.3.]
9. (Chebychev Inequality) Let I be a bounded interval and $f : I \to [0, \infty)$ be integrable over I. Let $r > 0$ and set $A = \{t \in I : f(t) > r\}$. Show that A is integrable and

$$\lambda(A) \leq 1/r \int_I f .$$

[Hint: Set $f_k = 1 \wedge k(f - r)^+$ and note $f_k \to C_A$, $0 \leq f_k \leq 1$.]

10. Let I be a bounded interval and $f : I \to \mathbb{R}$ absolutely integrable over I. Show that $\int_I |f| = 0$ if and only if $f(t) = 0$ for all $t \in I$ except those in a null subset. [Hint: $\{t : |f(t)| \neq 0\} = \bigcup_{k=1}^{\infty} \{t : |f(t)| > 1/k\}$; use Chebychev's Inequality.]
11. Let I be a closed interval in $\mathbb{R}^*$ and suppose $f : I \to \mathbb{R}$ is absolutely integrable over every bounded subinterval of I. Set $E = \{t \in I : f(t) \neq 0\}$. Show that E is measurable. [Hint: Let J be a bounded subinterval of I and set $f_k = (k|f|) \wedge C_J$ and note $f_k \uparrow C_{E \cap J}$.]
12. In Exercise 11 show that when $\int_F |f|$ exists, then $E \cap F$ is measurable.
13. Prove Theorem 12. [Hint: Amalgamate all of the null sets in the statement.]
14. State and prove a generalization of Theorem 5.4 in the spirit of Theorem 11.

Chapter 7

The Space of Gauge Integrable Functions

The space of gauge integrable functions is a vector space under the operations of pointwise addition and scalar multiplication, and the space also has a natural semi-norm, called the Alexiewicz norm. In this chapter we describe a few of the basic properties of the space and give references to some of the more technical properties. This chapter requires a basic knowledge of normed linear spaces and employs more advanced techniques (including several starred results from the text) than are required for the reading of the remainder of this text.

We treat the case when $I = [a, b], -\infty < a < b < \infty$. Let $\mathcal{HK}(I)$ be the space of all gauge integrable functions defined on I. Then $\mathcal{HK}(I)$ is a vector space under the usual operations of pointwise addition and scalar multiplication. There is a natural semi-norm on $\mathcal{HK}(I)$ originally defined by Alexiewicz ([A]) via $\|f\| = \sup\{|\int_a^t f| : a \le t \le b\}$ (Exercise 1). If $f = 0$ a.e. in I, then $\|f\| = 0$, and, conversely, if $\|f\| = 0$, it follows from Theorem 2 of Appendix 3 that $f = 0$ a.e. in I. Thus, if we identify functions which are a.e. in I, then $\mathcal{HK}(I)$ is a normed space under the Alexiewicz norm.

In contrast to the case of the space of Lebesgue integrable functions with the L^1-norm, the space $\mathcal{HK}(I)$ is not complete under the Alexiewicz norm.

Example 1. *Let $p : [0, 1] \to \mathbb{R}$ be continuous and nowhere differentiable with $p(0) = 0$ ([DS] 11.20). Pick a sequence of polynomials $\{p_k\}$ such that $p_k \to p$ uniformly and $p_k(0) = 0$ (Weierstrass Approximation Theorem ([DS] 18.8)). By the FTC $p_k(t) = \int_0^t p'_k$ for every $t \in [0, 1]$ so $\{p'_k\}$ is a Cauchy sequence in $\mathcal{HK}([0, 1])$ with respect to the Alexiewicz norm. If there exists $f \in \mathcal{HK}([0, 1])$ such that $\|p'_k - f\| \to 0$, then $p_k(t) = \int_0^t p'_k \to \int_0^t f$ uniformly for $t \in [0, 1]$*

so $p(t) = \int_0^t f$. Theorem 2 of Appendix 3 then implies that p is differentiable a.e. which is a contradiction. Hence, $\mathcal{HK}(I)$ is not complete.

We give a description of the completion of $\mathcal{HK}(I)$ as a subspace of distributions. For a distribution $T \in \mathcal{D}'(\mathbb{R})$, we write DT for its distributional derivative ([Sw2], § 26.4.12, [Tr] § 23). Let $\mathcal{F}(I) = \{T \in \mathcal{D}'(\mathbb{R}) : DT = f$, where $f : \mathbb{R} \to \mathbb{R}$ is continuous, $f = 0$ on $(-\infty, a]$ and $f(t) = f(b)$ for $t \geq b\}$. Note that if $f \in \mathcal{HK}(I)$, then f induces an element F of $\mathcal{F}(I)$ by $F(t) = \int_a^t f$, $a \leq t \leq b$, $F(t) = 0$ for $t < a$ and $F(t) = \int_a^b f$ for $t > b$ (Theorem 2 of Appendix 3 and Corollary 2.3); i.e., the map $f \to F$ imbeds $\mathcal{HK}(I)$ into $\mathcal{F}(I)$. We define a norm on $\mathcal{F}(I)$ by setting $\|T\| = \sup\{|f(t)| : a \leq t \leq b\}$. Now $\mathcal{F}(I)$ is complete under this norm and the norm restricted to $\mathcal{HK}(I)$ is just the Alexiewicz norm so $\mathcal{F}(I)$ is the completion of $\mathcal{HK}(I)$. For details, see [MO].

Since $\mathcal{HK}(I)$ is not complete, it is natural to ask if $\mathcal{HK}(I)$ is second category (or a Baire space; [DS] § 24). This is, however, not the case; $\mathcal{HK}(I)$ is first category in itself. To see this let $C_0(I)$ be the set of all continuous functions on I which vanish at a equipped with the sup-norm, $\|f\| = \sup\{|f(t)| : a \leq t \leq b\}$, $f \in C_0(I)$. We can identify $\mathcal{HK}(I)$ with the subspace $\{F : F(t) = \int_a^t f, a \leq t \leq b, f \in \mathcal{HK}(I)\} = Y$. By Theorem 2 of Appendix 3, each $F \in Y$ is differentiable a.e. in I so Y is first category in the complete space $C_0(I)$ ([DS] p. 293-294; [Boas] p. 61). By Exercise 5 this means that Y is first category in itself.

Although $\mathcal{HK}(I)$ is not second category, it does have the very pleasant property of being barrelled; barrelled spaces are important as they are spaces for which the Uniform Boundedness Principle and the Banach-Steinhaus Theorem hold (see [Sw2], § 24, [Tr] § 33). There are a number of proofs of this fact. We will give one below which requires only basic results about the gauge integral and then give references to other proofs in the Notes/Remarks section at the end of the chapter. We begin with the results for the gauge integral which are required.

Theorem 2. *Let $f_k \in \mathcal{HK}(I)$ and $\{J_k\}$ be a pairwise disjoint sequence of open subintervals of I. If $X = I \setminus \bigcup_{k=1}^\infty J_k$, assume that the boundary of $X, \partial X$, is countable with $\partial X = \{r_i : i \in \mathbb{N}\}$. Assume that $H(J) = \sum_{k=1}^\infty \int_J C_{J_k} f_k$ exists for every subinterval J of I and $\lim_{\ell(J)\to 0} H(J) = 0$. If $f = \sum_{k=1}^\infty C_{J_k} f_k$ [pointwise], then $f \in \mathcal{HK}(I)$ and $\int_J f = H(J)$ for every subinterval J.*

Proof: Let $\varepsilon > 0$. For each j let γ_j be a gauge on I such that $|S(f_j, \mathcal{D}) - \int_I f_j| < \varepsilon/2^j$ whenever $\mathcal{D} << \gamma_j$ and let $\delta_j > 0$ be such that $\ell(J) < \delta_j$ implies $|H(J)| < \varepsilon/2^j$. Define a gauge γ on I by $\gamma(t) = \gamma_j(t) \cap J_j$ for $t \in J_j$, $\gamma(t) = X^0$

for $t \in X^0$ and if $t = r_j$ choose $\gamma(t)$ to be a symmetric neighborhood of t with $\ell(\gamma(t)) < \delta_j$. Note the following properties relative to γ:

(i) if $t \in J_j$ and $t \in J \subset \gamma(t)$, then $f(t) = f_j(t)$ and $H(J) = \int_J f_j$;
(ii) if $t \in X^0$ and $t \in J \subset X^0$, then $f(t) = 0$ and $H(J) = 0$;
(iii) if $t = r_j$ and $r_j \in J \subset \gamma(r_j)$, then $f(t) = 0$ and $|H(J)| < \varepsilon/2^j$.

Suppose $\mathcal{D} = \{(t_i, I_i) : 1 \le i \le m\} << \gamma$. Set $d_i = \{k : t_k \in J_i\}$ for $i = 1, 2, \ldots$,

$$d_0 = \{k : t_k = r_j \text{ for some } j\},$$

$d_{-1} = \{k : t_k \in X^0\}$ and $\mathcal{D}_j = \{(t_i, I_i) : i \in d_j\}$ for $j = -1, 0, 1, \ldots$. Note
(iv) $\mathcal{D}_j << \gamma_j$ for $j = 1, 2, \ldots$.

Then

$$\text{(1)} \qquad \begin{aligned} |S(f,\mathcal{D}) - H(I)| &= \left|\sum_{i=-1}^{\infty} \sum_{k\in d_i} \{f(t_k)\ell(I_k) - H(I_k)\}\right| \\ &\le \left|\sum_{k\in d_{-1}} \{f(t_k)\ell(I_k) - H(I_k)\}\right| \\ &\quad + \left|\sum_{k\in d_0} \{f(t_k)\ell(I_k) - H(I_k)\}\right| \\ &\quad + \left|\sum_{i=1}^{\infty} \sum_{k\in d_i} \{f(t_k)\ell(I_k) - H(I_k)\}\right| \\ &= R_{-1} + R_0 + R_1 , \end{aligned}$$

with obvious definitions for R_i.

First, $R_{-1} = 0$ by (ii). Next, if $t_k = r_j$ for some j, then by (iii) $f(t_k) = 0$ and $\ell(I_k) < \delta_j$ so $|H(I_k)| < \varepsilon/2^j$, and since r_j is the tag for at most 2 subintervals in $\mathcal{D}$, $R_0 < 2\sum_{j=1}^{\infty} \varepsilon/2^j = 2\varepsilon$. Finally, if $t_k \in d_i$ for $i \ge 1$, then by (i) and (iv) and Henstock's Lemma,

$$\begin{aligned} \left|\sum_{k\in d_i} \{f(t_k)\ell(I_k) - H(I_k)\}\right| &= \left|\sum_{k\in d_i} \{f_i(t_k)\ell(I_k) - \int_{I_k} f_i\}\right| \\ &= \left|S(f_i, \mathcal{D}_i) - \int_{\underset{k\in d_i}{\cup I_k}} f_i\right| \le \varepsilon/2^i . \end{aligned}$$

Thus, $R_1 \le \varepsilon$ and (1) implies that $|S(f, \mathcal{D}) - H(I)| < 3\varepsilon$.

Hence, f is integrable over I with $\int_I f = H(I)$. The same computation works for every subinterval J; just begin with $\mathcal{D}$ being a γ-fine tagged partition of J.

For our next result we need another norm on $\mathcal{HK}(I)$ which is equivalent to the Alexiewicz norm. Let $\mathcal{I}$ be the family of all subintervals of I and set $\|f\|' = \sup\{|\int_J f| : J \in \mathcal{I}\}$ for $f \in \mathcal{HK}(I)$. It is easily seen that $\|f\| \le \|f\|' \le 2\|f\|$ for $f \in \mathcal{HK}(I)$ so $\|\ \|$ and $\|\ \|'$ are equivalent (Exercise 2).

Proposition 3. *Let $f_k \in \mathcal{HK}(I)$ with $\sum_{k=1}^{\infty} \|f_k\|' < \infty$ and let $\{J_k\}$ be a pairwise disjoint sequence of open subintervals of I. Set $f = \sum_{k=1}^{\infty} C_{J_k} f_k$ [pointwise].*

(i) *Then $H(J) = \sum_{k=1}^{\infty} \int_J C_{J_k} f_k$ converges uniformly for $J \in \mathcal{I}$.*
(ii) $\lim_{\ell(J)\to 0} H(J) = 0$.
(iii) *If $f \in \mathcal{HK}(I)$, then $\|f - \sum_{k=1}^{n} C_{J_k} f_k\|' \to 0$ [i.e., $\int_J f = \sum_{k=1}^{\infty} \int_J C_{J_k} f_k$ uniformly for $J \in \mathcal{I}$].*

Proof: Since $\|\int_J C_{J_k} f_k\| \le \|f_k\|'$ for every $J \in \mathcal{I}$, (i) and (iii) follow immediately.

For (ii), let $\varepsilon > 0$. Choose n such that $\sum_{k=n+1}^{\infty} \|f_k\|' < \varepsilon/2$ and choose $\delta > 0$ such that $\ell(J) < \delta$ implies $|\int_J C_{J_k} f_k| < \varepsilon/(2n)$ for $k = 1, \dots, n$ [this follows from the uniform continuity of the definite integral $F(t) = \int_a^t f$ of a gauge integrable function, Corollary 3.2]. Then $\ell(J) < \delta$ implies $\|H(J)\| < \varepsilon$.

To establish the barrelledness of $\mathcal{HK}(I)$ we require the following result called the Antosik-Mikusinski Matrix Theorem.

Theorem 4. *Let $a_{ij} \in \mathbb{R}$ for $i, j \in \mathbb{N}$ and $A = [a_{ij}]$. Suppose*

(i) $\lim_i a_{ij} = 0$ *for every j and*
(ii) *for every increasing sequence of positive integers $\{m_j\}$ there is a subsequence $\{n_j\}$ of $\{m_j\}$ such that $\lim_i \sum_{j=1}^{\infty} a_{in_j}$ exists.*

Then $\lim_i a_{ij} = 0$ uniformly for $j \in \mathbb{N}$. In particular, $\lim_i a_{ii} = 0$.

For a proof (of a more general result) see [Sw1] 2.8.2 or [Sw2] 9.2.

A normed space Y is barrelled if every weak* bounded subset of the dual space Y' is norm bounded ([Sw2] 24.5, [Tr] 33.1). We use this dual space characterization of barrelled normed spaces, but we do not require a concrete description of the dual of $\mathcal{HK}(I)$.

Theorem 5. *Let $B \subset \mathcal{HK}(I)'$ be weak* bounded. Then B is norm bounded [i.e., $\mathcal{HK}(I)$ is barrelled].*

Proof: Let $A \subset \mathcal{HK}(I)$ be bounded with $\alpha = \sup\{\|f\| : f \in A\}$. It suffices to show that $\beta = \sup\{|\langle \nu, f\rangle| : \nu \in B,\ f \in A\} < \infty$. Suppose $\beta = \infty$. Then there exist $\nu_1 \in B$, $f_1 \in A$ such that $|\langle \nu_1, f_1\rangle| > 2$. By Exercise 4, $t \to \langle \nu_1, C_{[a,t]} f_1\rangle$ is continuous so there exists a_1 such that $|\langle \nu_1, C_{[a,a_1]} f_1\rangle| > 1$ and $|\langle \nu_1, C_{(a_1,b]} f_1\rangle| > 1$. Either $[a, a_1]$ or $(a_1, b]$ satisfies $\sup\{|\langle \nu, C_{[a,a_1]} f\rangle| : \nu \in B, f \in A\} = \infty$ or $\sup\{|\langle \nu, C_{(a_1,b]} f\rangle| : \nu \in B, f \in A\} = \infty$. Pick one of these intervals which satisfies this condition and label it I_1 and put $J_1 = I \setminus I_1$. Note $|\langle \nu_1, C_{J_1} f_1\rangle| > 1$. Now there exist $\nu_2 \in B, f_2 \in A$ such that $|\langle \nu_2, C_{I_1} f_2\rangle| > 2^4$. By the argument above there is a partition of I_1 into 2 disjoint subintervals A_2, B_2 such that $|\langle \nu_2, C_{A_2} f_2\rangle| > 2^3, |\langle \nu_2, C_{B_2} f_2\rangle| > 2^3$ and $\sup\{|\langle \nu, C_{A_2} f\rangle| : \nu \in B, f \in A\} = \infty$. Put $J_2 = B_2, I_2 = A_2$ and continue this construction. This produces a pairwise disjoint sequence of intervals $\{J_j\}, \{f_j\} \subset A, \{\nu_j\} \subset B$ such that

(2) $|\langle \nu_j, C_{J_j} f_j\rangle| > j^3$.

Let K_j be the interior of J_j; then (2) holds with J_j replaced by K_j. Set $h_j = \frac{1}{j^2} f_j$. Then $\|h_j\| \le \frac{1}{j^2}\|f_j\| \le \alpha/j^2$. Hence $\sum_{j=1}^{\infty} \|h_j\| < \infty$.

Consider the matrix $M = [m_{ij}] = [\frac{1}{i}\langle \nu_i, C_{K_j} h_j]$. We claim that M satisfies (I) and (II) in Theorem 4. First, (I) holds by the weak* boundedness of B. Next, Theorem 2 and Proposition 3 are applicable to the sequences $\{K_j\}$ and $\{h_j\}$ since if $X = I \setminus \bigcup_{j=1}^{\infty} K_j$, the boundary of X consists of the endpoints of the K_j and possibly 2 other limit points. Therefore, if $h = \sum_{j=1}^{\infty} C_{K_j} h_j$ [pointwise], then $h \in \mathcal{HK}(I)$ and the series converges in the norm of $\mathcal{HK}(I)$. By the continuity of each ν_i, $\sum_{j=1}^{\infty} m_{ij} = \langle \frac{1}{i}\nu_i, h\rangle$ and $\langle \frac{1}{i}\nu_i, h\rangle \to 0$ by the weak* boundedness of B. Since the same argument can be applied to any subsequence of $\{h_j\}$, condition (II) is satisfied. Theorem 4 implies $m_{ii} \to 0$ contradicting (2).

Thus, $\mathcal{HK}(I)$ supplies another example of a first category, barrelled normed space (see [Sw2], Exer. 24.6, for another, more familiar, example).

We give a description of the dual of $\mathcal{HK}(I)$.

Theorem 6. *$F \in \mathcal{HK}(I)'$ if and only if there is a right continuous function f of bounded variation with $f(b) = 0$ such that $\langle F, \varphi\rangle = \int_I \varphi(t) f(t)dt$ for every $\varphi \in \mathcal{HK}(I)$. Moreover, $\|F\| = Var(f : I)$.*

See [A], Theorem 1 for a proof. For references to other description of the dual, see the Notes/Remarks section.

We show that the step functions are dense in $\mathcal{HK}(I)$ (a step function is a linear combination of characteristic functions of intervals).

Theorem 7. *The step functions are dense in $\mathcal{HK}(I)$.*

Proof: Let $f \in \mathcal{HK}(I)$ and $\varepsilon > 0$. Pick a gauge γ on I such that $|S(f,\mathcal{D}) - \int_I f| < \varepsilon$ whenever $\mathcal{D} << \gamma$. Fix $\mathcal{D} = \{(t_i, I_i) : 1 \le i \le m\} << \gamma$. Set $\varphi = \sum_{i=1}^m f(t_i) C_{I_i}$. By the Uniform Henstock Lemma (3.8) if $J \in \mathcal{I}$,

$$\left|\int_J f - \int_J \varphi\right| = \left|\sum_{i=1}^m \left\{\int_{J\cap I_i} f - \int_{J\cap I_i} \varphi\right\}\right|$$
$$= \left|\sum_{i=1}^m \left\{\int_{J\cap I_i} f - f(t_i)\ell(I_i \cap J)\right\}\right| \le 3\varepsilon .$$

Hence, $\|f - \varphi\|' \le 3\varepsilon$.

Finally, we establish a sufficient condition for convergence of a sequence with respect to the Alexiewicz norm.

Theorem 8. *If $\{f_k\}$ is uniformly integrable and $\{f_k\}$ convergence pointwise to f, then $\{f_k\}$ converges to f in norm.*

Proof: By Theorem 5.2 f is integrable so we may assume that $f = 0$. Let $\varepsilon > 0$. Let γ be a gauge such that $|S(f_k, \mathcal{D}) - \int_I f_k| < \varepsilon$ for every k whenever $\mathcal{D} << \gamma$. Fix a tagged partition $\mathcal{D} = \{(t_i, I_i) : 1 \le i \le m\}$ which is γ-fine. Choose n such that $k \ge n$ implies $|f_k(t_i)| < \varepsilon / m\ell(I_i)$ for $i = 1, \ldots, m$. Suppose that J is an arbitrary subinterval of I and $k \ge n$. From the Uniform Henstock Lemma (3.8), we have

$$\left|\int_J f_k\right| \le \sum_{i=1}^m \left|\int_{J\cap I_i} f_k - f_k(t_i)\ell(I_i \cap J)\right| + \sum_{i=1}^m |f_k(t_i)|\ell(I_i \cap J)$$
$$< 6\varepsilon + \varepsilon = 7\varepsilon .$$

Hence, if $k \ge n$, then $\|f_k\| \le 7\varepsilon$.

It follows from Theorem 8 that the convergence conclusions in the MCT and DCT can be improved to norm convergence, i.e., $\lim \int_J f_k = \int_J f$ uniformly for all subintervals J.

Notes/Remarks

For another example showing that $\mathcal{HK}(I)$ is not complete, see [LPY] 11.1.

Sargent first showed that the space $\mathcal{HK}(I)$ is barrelled. Actually Sargent showed that the space of Denjoy integrable functions on I is barrelled under the Alexiewicz norm ([Sar1]); since the Denjoy integral is equivalent to the gauge integral ([Go1]), it follows that $\mathcal{HK}(I)$ is barrelled. Sargent's proof used a characterization of the dual space which is not required in the proof of Theorem 5. Thomson has given another proof of the barrelledness of $\mathcal{HK}(I)$ which does not use the dual characterization of barrelledness ([Th2]). See also [Gi] for another proof of barrelledness (and actually a stronger topological property).

The space $\mathcal{HK}(I)$ has a stronger property than being barrelled which is also of interest. A topological vector space X is called a *Sargent space* (β-space by Sargent) if there exists no sequence of subsets $\{E_k\}$ satisfying

(S1) $0 \in E_k$, $E_k - E_k \subset E_{k+1}$
(S2) $X = \bigcup_{k=1}^{\infty} E_k$
(S3) Every E_k is nowhere dense.

A second category, metrizable space is clearly a Sargent space, but $\mathcal{HK}(I)$ furnishes an example of a first category Sargent space ([Sar2]; see also [LPY] 11.3). Sargent spaces are important because the Uniform Boundedness Principle and the Banach-Steinhaus Theorem hold for such spaces ([Sar2], [Sw2] 9.14-15, [Tr] § 33).

For other descriptions of the dual of $\mathcal{HK}(I)$ other than that given in Theorem 6, see [Sar1], Theorem 4, [LPY], Theorems 12.3 and 12.7 and [MO] 2.1. In particular, Sargent describes the dual of $\mathcal{HK}(I)$ as the space $EBV(I)$ of functions which are equivalent to a function of bounded variation; i.e., $f \in EBV(I)$ if and only if there is a function $g : I \to \mathbb{R}$ of bounded variation on I such that $f = g$ a.e. in I. A function f on I is a multiplier for $\mathcal{HK}(I)$ if and only if fg is integrable over I for every $g \in \mathcal{HK}(I)$. (See Notes/Remarks in Chapter 4.) Sargent shows that f is a multiplier if and only if $f \in EBV(I)$ so the dual space and the space of multipliers coincide ([Sar3]; [LPY] 12.9). For the higher dimensional case see [LCL1].

When I is an unbounded interval let $\mathcal{HK}(I)$ be the space of all gauge-integrable functions defined on I. If $I = [-\infty, \infty]$, the (Alexiewicz) norm on $\mathcal{HK}(I)$ is defined to be $\|f\| = \sup\{|\int_{-\infty}^{t} f| : -\infty < t < \infty\}$. If $\mathcal{I}$ is the family of all subintervals of I, then a norm equivalent to the Alexiewicz norm is given by $\|f\|' = \sup\{|\int_J f| : J \in \mathcal{I}\}$. [Similar definitions are made in case $I = [-\infty, a]$ or $I = [a, \infty]$.] The completion of $\mathcal{HK}(I)$ can be described as above. The barrelledness of $\mathcal{HK}(I)$ in the case when I is unbounded can be established as in Theorem 5. However, there does not seem to be a concrete description of the dual space of $\mathcal{HK}(I)$ in the case of an unbounded interval.

Finally, Theorem 7 should be compared with Theorem 5.10 of [LPY] where it is shown that every gauge integrable function is the a.e. pointwise limit of step functions.

Exercises

1. Show that the Alexiewicz norm satisfies the axioms of a semi-norm, i.e., $\|f+g\| \le \|f\| + \|g\|$ and $\|tf\| = |t|\|f\|$.
2. Show that the expression for $\| \ \|'$ following Theorem 2 defines a semi-norm on $\mathcal{HK}(I)$ with $\|f\| \le \|f\|' \le 2\|f\|$.
3. Show that $\mathcal{HK}(I)$ is separable.
4. Let $f \in \mathcal{HK}(I)$. Define $\hat{f} : I \to \mathcal{HK}(I)$ by $\hat{f}(t) = C_{[a,t]}f$. Show $\hat{f}$ is continuous.
5. Let Y be a dense subspace of the normed linear space X. Show that the category of Y in X is the same as the category of Y in itself. ([AM] p. 201).
6. Let $g \in BV(I)$. Use Theorem 4.15 to show that $\langle G, f\rangle = \int_I fg$ defines a continuous linear functional G on $\mathcal{HK}(I)$ with $\|G\| \le Var(g : I) + |g(b)|$.

Chapter 8

Multiple Integrals and Fubini's Theorem

Except for the usual cumbersome notation it is straightforward to define the gauge integral for functions defined on intervals in n-dimensional Euclidean space, $\mathbb{R}^n$. In this chapter we make such a definition and study some of the basic properties of this multiple integral in $\mathbb{R}^n$. Most of the proofs in 1 dimension carry forward to higher dimensions except for the usual notational complications; we do not repeat the proofs which are obvious modifications of earlier proofs in 1 dimension.

The computation of integrals over subsets of $\mathbb{R}^n$ (multiple integrals) is usually carried out by doing iterated 1-dimensional integrals. Theorems which relate multiple integrals to iterated integrals are usually referred to as Fubini Theorems. We give 2 versions of Fubini's Theorem. The first version (junior grade) uses only basic properties of the gauge integral and should suffice for an introductory real analysis course. The second version is more general but uses results relating null sets and the gauge integral from Chapter 6 on Lebesgue measure; it is included to show the generality of the gauge integral in $\mathbb{R}^n$ for those readers who have gone through Chapter 6 [at least the part up to Corollary 6.6].

Let $\mathbb{R}^{*n} = \mathbb{R}^* \times \cdots \times \mathbb{R}^*$ where there are n factors in the product. An *interval* I in $\mathbb{R}^{*n}$ is a product $I_1 \times \cdots \times I_n$ where each I_k is an interval in $\mathbb{R}^*$; we say that I is a closed (open) interval if each I_k is closed (open). The *interior* of I, written I°, is the product of the interiors of the I_k. The *volume* of an interval I, denoted $v(I)$, is the product $v(I) = \prod_{k=1}^n \ell(I_k)$, where we continue to use the convention that $0 \cdot \infty = 0$. [In $\mathbb{R}^2$, the term *area* would be more appropriate.]

If I is a closed interval, a *partition* of I is a finite collection of closed subintervals of I, $\{I_k : k = 1, \dots, m\}$, such that $I_k^\circ \cap I_j^\circ = \emptyset$ if $k \neq j$ [i.e., the $\{I_k\}$ do not overlap] and $I = \bigcup_{k=1}^m I_k$. A *tagged partition* $\mathcal{D}$ of I is a finite collection of pairs $\{(x_i, I_i) : 1 \leq i \leq m\}$ such that $\{I_i : 1 \leq i \leq m\}$ is a partition of I and $x_i \in I_i$ for $1 \leq i \leq m$; we call the I_i subintervals of $\mathcal{D}$ and call x_i the *tag* associated with I_i. A *gauge* γ on I is a function defined on I such that $\gamma(x)$ is an open interval containing x for every $x \in I$. A tagged partition $\mathcal{D} = \{(x_i, I_i) : 1 \leq i \leq m\}$ is γ-*fine* if $x_i \in I_i \subset \gamma(x_i)$ for $1 \leq i \leq m$; we write $\mathcal{D} << \gamma$ as before.

We next proceed to show that every gauge γ defined on a closed interval in $\mathbb{R}^{*n}$ has a γ-fine tagged partition. For simplicity and ease of understanding, we write out most of the proofs for $n = 2$; the proofs can be adapted to $\mathbb{R}^{*n}$ with the usual notational complications. We give a proof of the existence of γ-fine tagged partitions based on their existence in $\mathbb{R}$; the construction is used later in the proof of Fubini's Theorem 13 [a more straightforward proof for compact intervals is outlined in Exercise 1].

Let J and K be closed intervals in $\mathbb{R}^*$ and $I = J \times K$. Let γ be a gauge on I. If P_1 and P_2 are the projections from $\mathbb{R}^{*2}$ onto the first and second coordinates, respectively, then

$$\gamma_1(x, y) = P_1\gamma(x, y)(\gamma_2(x, y) = P_2\gamma(x, y))$$

has the property that $\gamma_1(\cdot, y)(\gamma_2(x, \cdot))$ defines a gauge on $J(K)$ for each $y \in K(x \in J)$ [Draw a picture!].

Lemma 1. *For $y \in K$ let $\mathcal{P}_y$ be a $\gamma_1(\cdot, y)$-fine tagged partition of J.*

(i) *If $\gamma_2'(y) = \cap\{\gamma_2(x, y) : (x, A) \in \mathcal{P}_y\}$, then γ_2' is a gauge on K.*

(ii) *If $\mathcal{K} = \{(y_j, K_j) : 1 \leq j \leq m\}$ is a γ_2'-fine tagged partition of K and if*

$$\mathcal{P}_y = \{(p_j^y, P_j^y) : 1 \leq j \leq p_y\},$$

then

$$\mathcal{D} = \{((p_i^{y_j}, y_j), P_i^{y_j} \times K_j) : 1 \leq j \leq m, 1 \leq i \leq p_{y_j}\}$$

is a γ-fine tagged partition of I. $\mathcal{D}$ is called the compound tagged partition generated by $\{\mathcal{P}_y : y \in K\}$ and $\mathcal{K}$. [See Figure 1.]

Proof: (i) is clear since the intersection in (i) is finite.

For (ii), if $(p_i^{y_j}, y_j) \in P_i^{y_j} \times K_j$, then $y_j \in K_j \subset \gamma_2'(y_j) \subset \gamma_2(p_i^{y_j}, y_j)$ and

$$p_i^{y_j} \in P_i^{y_j} \subset \gamma_1(p_i^{y_j}, y_j).$$

Hence,

$$P_i^{y_j} \times K_j \subset \gamma_1(p_i^{y_j}, y_j) \times \gamma_2(p_i^{y_j}, y_j) = \gamma(p_i^{y_j}, y_j)$$

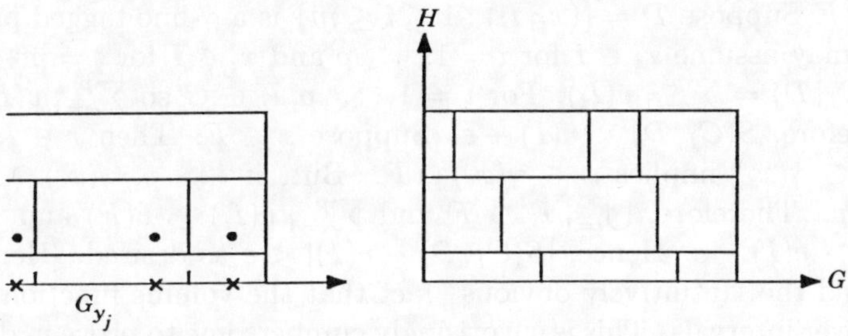

Figure 8.1

Corollary 2. *Every gauge γ on I has a γ-fine tagged partition.*

If $\mathcal{D} = \{(x_i, I_i) : 1 \le i \le m\}$ is a tagged partition of a closed interval I and $f : I \to \mathbb{R}$, the *Riemann sum* of f with respect to $\mathcal{D}$ is defined to be $S(f, \mathcal{D}) = \sum_{i=1}^m f(x_i)v(I_i)$. Here we are assuming, as was the case in $\mathbb{R}^*$, that all functions vanish at infinite points and $0 \cdot \infty = 0$.

Definition 3. *Let I be a closed interval in $\mathbb{R}^{*n}$ and $f : I \to \mathbb{R}$. Then f is (gauge) integrable over I if there exists $A \in \mathbb{R}$ such that for every $\varepsilon > 0$ there is a gauge γ on I such that $|S(f, \mathcal{D}) - A| < \varepsilon$ whenever $\mathcal{D} << \gamma$.*

From Corollary 2, Definition 3 is meaningful and the uniqueness of A follows as in Theorem 1.4. A is called the integral of f over I and is denoted by $\int_I f = \int_I f(x)dx$.

The basic properties of the integral such as linearity, additivity, positivity, etc. carry forward to $\mathbb{R}^{*n}$ as before; we do not repeat their statements or proofs. There is one difference between the integral in $\mathbb{R}^{*n}$ and $\mathbb{R}^*$ that should be noted. A tag in a tagged partition can be the tag for 2^n subintervals for intervals in $\mathbb{R}^{*n}$; this can lead to some slightly different estimates for Riemann sums.

We give a simple example illustrating the integral in $\mathbb{R}^2$. If I is an interval in $\mathbb{R}^{*n}$, its closure, written $\bar{I}$, is the product of the closures of the component intervals defining I.

Example 4. *Let I be a bounded interval in $\mathbb{R}^n$. Then C_I is integrable over I with $\int_{\mathbb{R}^n} C_I = v(I)$.*

We sketch the proof for $n = 2$. Let $\varepsilon > 0$. Pick a closed interval $F \subset I$ and an open interval $G \supset I$ such that $v(G) - v(F) < \varepsilon$. Define a gauge on $\mathbb{R}^2$ by setting $\gamma(x) = G$ when $x \in I$ and choosing $\gamma(x)$ such that $\gamma(x) \cap F = \emptyset$ when $x \notin I$. Suppose $\mathcal{D} = \{(x_i, I_i) : 1 \le i \le m\}$ is a γ-fine tagged partition of $\mathbb{R}^{*2}$. We may assume $x_i \in I$ for $i = 1, \ldots, p$ and $x_i \notin I$ for $i = p+1, \ldots, m$. Then $S(C_I, \mathcal{D}) = \sum_{i=1}^p v(I_i)$. For $i = 1, \ldots, p, I_i \subset G$ so $\sum_{i=1}^p v(I_i) \le v(G)$ and, therefore, $S(C_I, \mathcal{D}) < v(I) + \varepsilon$. Suppose $x \in F$. Then $x \in I_i$ for some i and $I_i \subset \gamma(x_i)$ implies $x \in \gamma(x_i) \cap F$. But, if $i > p$, $\gamma(x_i) \cap F = \emptyset$ so $1 \le i \le p$. Therefore, $\bigcup_{i=1}^p I_i \supset F$ and $\sum_{i=1}^p v(I_i) \ge v(F)$ and, therefore, $S(C_I, \mathcal{D}) > v(I) - \varepsilon$. Hence, $|S(C_I, \mathcal{D}) - v(I)| < \varepsilon$ as desired. [Here we have tacitly used the "intuitively obvious" fact that the volume function is finitely additive over intervals. This is surprisingly cumbersome to prove in dimensions larger than 1. See for example [MS2] IV.2.1.]

It follows from Example 4 and linearity of the integral that if φ is a *step function*, a linear combination of characteristic functions of bounded intervals, then φ is integrable over $\mathbb{R}^{*n}$ and its integral is given in a natural way.

We next give a further example which shows that the gauge (double) integral in higher dimensions also admits conditionally convergent integrals (Exercise 6).

Example 5. *Let $\sum_{k=1}^\infty a_k$ be a convergent real series and $I = [0,1] \times [0,1]$. Set*

$$K_j = [1/2^j, 1/2^{j-1}], L_j = K_j \times K_j \text{ and } J_j = L_{2j}\,.$$

Then $\{J_j\}$ is a pairwise disjoint sequence of closed subintervals of I with $v(J_j) = 1/2^{4j}$. Let $0 < \varepsilon < 1$. Pick a pairwise disjoint sequence $\{O_j\}$ of open subintervals of I such that $O_j \supset J_j$ and

$$v(J_j) + \varepsilon/2^{5j} > v(O_j) > v(J_j)$$

[draw a picture!]. Put $f = \sum_{j=1}^\infty a_j 2^{4j} C_{J_j}$. We show that f is integrable over I with $\int_I f = \sum_{j=1}^\infty a_j$. Pick N such that $|\sum_{i=k}^\infty a_i| < \varepsilon$ and $|a_k| < \varepsilon$ for $k \ge N$ and let $M = \sup\{|a_k| : k \in \mathbb{N}\}$. We now define a gauge γ on I. If $x \in J_j$, set $\gamma(x) = O_j$; if $x \notin J_j$ for any j and $x \ne 0$, pick $\gamma(x)$ to be an open interval containing x such that $0 \notin \gamma(x)$ and $\gamma(x) \cap J_j = \emptyset$ for every j; for $x = 0$, let $\gamma(0) = (-\delta, \delta) \times (-\delta, \delta)$ where $\gamma(0) \supset J_k$ for $k \ge N+1$ and $\gamma(0) \cap J_N = \emptyset$.

Now suppose that $\mathcal{D} = \{(x_i, I_i) : 0 \le i \le m\}$ is γ-fine with $0 \in I_0$; note $x_0 = 0$. Let q be the largest integer such that J_q is not contained in the interior *of I_0; note that $q \geq N$ since $I_0 \subset \gamma(0)$ and $\gamma(0)$ contains all J_k for $k \geq N+1$. Note also that*

$$S(f,\mathcal{D}) = \sum_{i=1}^{m} f(x_i)v(I_i) = \sum_{k=1}^{q} \sum_{x_j \in J_k} f(x_j)v(I_j)$$

$$= \sum_{k=1}^{q} a_k 2^{4k} \sum_{x_j \in J_k} v(I_j)$$

since $f(x) \neq 0$ if and only if $x \in J_k$ and $a_k \neq 0$ for some k. If $k < q$, then

$$v(J_k) + \varepsilon/2^{5k} = 1/2^{4k} + \varepsilon/2^{5k} > v(O_k) > v(J_k) = 1/2^{4k}$$

and

$$v(O_k) \geq \sum_{x_j \in J_k} v(I_j) \geq v(J_k) = 1/2^{4k}$$

so

(1) $\varepsilon/2^k \geq 2^{4k} \sum_{x_j \in J_k} v(J_j) - 1 \geq 0$ *for $k < q$.*

For $k = q$,

$$\sum_{x_j \in J_q} v(I_j) \leq v(O_q) < 1/2^{4q} + \varepsilon/2^{5q} < 2/2^{4q}$$

so

(2) $\left| \sum_{x_j \in J_q} f(x_j)v(I_j) \right| = |a_q| 2^{4q} \sum_{x_j \in J_q} v(I_j) \leq 2|a_q| < 2\varepsilon$

since $q \geq N$. From (1) and (2), we have

$$\left| S(f,\mathcal{D}) - \sum_{k=1}^{\infty} a_k \right| \leq \left| \sum_{k=1}^{q-1} a_k \left(2^{4k} \sum_{x_j \in J_k} v(I_j) - 1 \right) \right|$$

$$+ \left| a_q 2^{4q} \sum_{x_j \in J_q} v(I_j) \right| + |a_q| + \left| \sum_{k=q+1}^{\infty} a_k \right|$$

$$\leq \sum_{k=1}^{q-1} |a_k| \varepsilon/2^k + 2\varepsilon + \varepsilon + \varepsilon < M\varepsilon + 4\varepsilon = (M+4)\varepsilon\,.$$

Hence, $\int_I f = \sum_{k=1}^{\infty} a_k$.

We leave it to Exercise 6 to show that the function f is absolutely integrable if and only if the series $\sum a_k$ is absolutely convergent.

We next establish the n-dimensional analogue of Theorem 2.10 on the integrability of continuous functions. We first require a preliminary result.

Lemma 6. *Let* $K \subset \mathbb{R}^n$ *be compact and* I *a compact interval containing* K*. If* $f : K \to \mathbb{R}$ *is continuous and non-negative on* K *and if* f *is extended to* I *by setting* $f = 0$ *on* $I \setminus K$*, then there is a sequence of non-negative step functions* $\{s_k\}$ *vanishing outside* I *such that* $s_k(x) \downarrow f(x) \,\forall\, x \in I$*, where the convergence is uniform for* $x \in K$*.*

Proof: For each $k \in \mathbb{N}$ divide I into a finite collection $\mathcal{F}_k$ of pairwise disjoint intervals such that the diameter of each subinterval in $\mathcal{F}_k$ is smaller than $1/k$. Make the choice such that $\mathcal{F}_{k+1}$ is a refinement of $\mathcal{F}_k$, that is, each subinterval in $\mathcal{F}_{k+1}$ is contained in some interval of $\mathcal{F}_k$.

Construct s_k from $\mathcal{F}_k$ as follows: for $x \in I$ there is a unique $J \in \mathcal{F}_k$; set $s_k(x) = \sup\{f(x) : x \in J\}$ and $s_k(x) = 0$ for $x \notin I$. Note $s_k(x) \downarrow$ since $\mathcal{F}_{k+1}$ is a refinement of $\mathcal{F}_k$.

We claim that $s_k(x) \downarrow f(x)$: First, suppose that $x \in I \setminus K$. Then, there is an open interval I_x containing x and such that $I_x \cap K = \emptyset$. Any interval of sufficiently small diameter containing x will be contained in I_x. Since f is zero outside K, this implies that $s_k(x) = 0$ for large k so $s_k(x) \downarrow f(x)$. For $x \in K$, let $\varepsilon > 0$. There is an open interval I_x containing x such that $|f(x) - f(y)| < \varepsilon$ for $y \in K \cap I_x$. Thus, $0 \le f(x)$ and $f(y) \le f(x) + \varepsilon$ for all $y \in I_x$ since f is zero outside K. For large k, the interval of $\mathcal{F}_k$ which contains x will be a subset of I_x so $s_k(x) \le f(x) + \varepsilon$ for large k. Since $f(x) \le s_k(x)$ for all x, it follows that $s_k(x) \downarrow f(x)$.

The uniform convergence on K follows from the uniform continuity of f on K, that is, the estimate $s_k(x) \le f(x) + \varepsilon$ holds uniformly for $x \in K$ and large k.

Note that when the function f is extended from K to I, the extended function will not, in general, be continuous on I or even K.

If f is not non-negative, we may decompose $f = f^+ - f^-$ and apply Lemma 6 to both f^+ and f^-. We then obtain a sequence of step functions vanishing outside I, $\{s_k\}$, which converges pointwise to f on I and $s_k \to f$ uniformly on K; however, the convergence is now not monotone.

Theorem 7. *Let $K \subset \mathbb{R}^n$ be compact and $f : K \to \mathbb{R}$ continuous on K. If f is extended to $\mathbb{R}^{*n}$ by setting $f = 0$ outside of K, then f is integrable over $\mathbb{R}^{*n}$.*

Proof: By considering $f = f^+ - f^-$, we may assume that f is non-negative. Choose a compact interval I containing K and let $\{s_k\}$ be the sequence of step functions converging to f given in Lemma 6. Each s_k vanishes outside I so each is integrable. Since $s_k \downarrow f$ and $\int_I s_k \geq 0$ for every k, the MCT implies that f is integrable over $\mathbb{R}^{*n}$.

Theorem 7 gives a reasonably broad class of integrable functions which are sufficient for many of the applications encountered in introductory real analysis courses.

FUBINI'S THEOREM: JUNIOR GRADE

In this section we establish a version of Fubini's Theorem which is adequate for most applications encountered in an introductory real analysis course. Again we establish the results in $\mathbb{R}^{*2}$, but the methods work equally well in $\mathbb{R}^{*n}$.

Let J, K be closed intervals in $\mathbb{R}^*$, $I = J \times K$ and $f : I \to \mathbb{R}$. The usual method for evaluating (double) integrals over I, $\int_I f$, is to evaluate an iterated integral, $\int_K \int_J f(x,y)dxdy$. A theorem which asserts the equality of a double integral and an iterated integral is often referred to as a "Fubini Theorem". For a Fubini Theorem to hold we must establish the existence of the integral $\int_J f(x,y)dx$ for each y and then show the function $y \to \int_J f(x,y)dx$ is integrable over K with integral equal to $\int_I f$. Our next result gives conditions sufficient for this to hold.

Theorem 8. *Suppose J and K are bounded closed intervals and $f : I \to \mathbb{R}$ is bounded. Suppose there exists a sequence of step functions $\{s_k\}$ such that $s_k \to f$ pointwise on I. Then*
(i) *for each $y \in K$, $f(\cdot, y)$ is integrable over J,*
(ii) *the function $F(y) = \int_J f(x,y)dx$ is integrable over K, and*
(iii) *the function f is integrable over I with*

$$\int_I f = \int_K \int_J f(x,y)dxdy\,.$$

Proof: First, suppose that f is a characteristic function of an interval, $f = C_{A\times B}$, A and B intervals in $\mathbb{R}$. For (i) and (ii) we have $\int_J f(\cdot, y) = \int_J C_A C_B(y) = C_B(y)\ell(A \cap J)$ and

$$\int_K \int_J f(\cdot, y)dy = \int_K C_B(y)\ell(A \cap J)dy = \ell(B \cap K)\ell(A \cap J)\,.$$

Since $\int_J f = v(I \cap A \times B) = \ell(B \cap K)\ell(A \cap J)$, (iii) follows.

The validity of the theorem for step functions now follows immediately from the linearity of the integral.

Now consider the general case. Let $M > 0$ be such that $|f(z)| \le M$ for $z \in I$. For $z \in I$, set

$$\bar{s}_k(z) = \begin{cases} M & s_k(z) > M \\ s_k(z) & -M \le s_k(z) \le M \\ -M & s_k(z) < -M \,. \end{cases}$$

Then $\{\bar{s}_k\}$ is a sequence of step functions which converges pointwise to f on I and is uniformly bounded by M. By the first part of the proof, $\bar{s}_k(\cdot, y)$ is integrable over J for every $y \in K$ and $\{\bar{s}_k(\cdot, y)\}$ converges pointwise to $f(\cdot, y)$ on J. By the Bounded Convergence Theorem (Exercise 5.6), $f(\cdot, y)$ is integrable over J with $\lim \int_J \bar{s}_k(\cdot, y) = \int_J f(\cdot, y)$. Also, by the first part, each function $y \to \int_J \bar{s}_k(\cdot, y)$ is integrable over K, and we have just shown that this sequence of functions converges pointwise to the function $y \to \int_J f(\cdot, y)$. Since $|\int_J f(\cdot, y)| \le M\ell(J)$, the Bounded Convergence Theorem implies that $y \to \int_J f(\cdot, y)$ is integrable over K with $\lim \int_K \int_J \bar{s}_k(\cdot, y)dy = \int_K \int_J f(\cdot, y)dy$. Since $\{\bar{s}_k\}$ converges to f pointwise on I and the sequence is uniformly bounded by M, another application of the Bounded Convergence Theorem and the first part of the proof give

$$\lim \int_I \bar{s}_k = \lim \int_K \int_J \bar{s}_k(x, y)dxdy = \int_I f = \int_K \int_J f(\cdot, y)dy \,,$$

and the result is established.

Of course, by symmetry, Theorem 8 is also applicable to the other iterated integral, and, in particular, implies that the 2 iterated integrals $\int_K \int_J f(x, y)dxdy$ and $\int_J \int_K f(x, y)dydx$ are equal. This equality does not hold in general [see Exercise 2]. It is also the case that both iterated integrals may exist and are equal but the function may fail to be integrable [Exercise 3].

Note also that Lemma 6 gives a sufficient condition for the hypothesis in Theorem 8 to be satisfied.

We next establish a result which removes the boundedness assumptions from Theorem 8. The result also establishes a sufficient condition for integrability in terms of an iterated integral; such results are sometimes referred to as a "Tonelli Theorem".

Theorem 9. *Let $g : I \to \mathbb{R}$ be such that $|f(z)| \le g(z)$ for all $z \in I$ and the iterated integral $\int_K \int_J g(x,y)dxdy$ exists. Assume there exists a sequence of step functions, $\{s_k\}$, such that $s_k \to f$ pointwise on I. Then*

(i) *for each $y \in K$, $\int_J f(x,y)dx$ exists,*
(ii) *the function $F(y) = \int_J f(x,y)dx$ is integrable over K, and*
(iii) *f is integrable over I with*

$$\int_I f = \int_K \int_J f(x,y)dxdy\,.$$

Proof: If f satisfies the hypothesis of the theorem, then so do the functions f^+ and f^- so we may assume that f is non-negative.

For each k set $I_k = \{(x,y) : |x| \le k, |y| \le k\} \cap I$ and define f_k by $f_k = f \wedge (kC_{I_k})$; that is, f_k is altered by setting $f_k = 0$ outside I_k and truncating f at k [draw a picture]. Each f_k satisfies the hypothesis of Theorem 8 over the interval I_k. If $J_k = J \cap [-k,k]$ and $K_k = K \cap [-k,k]$, then Theorem 8 gives $\int_{I_k} f_k = \int_{K_k} \int_{J_k} f_k(x,y)dxdy$. Since f_k vanishes outside I_k, we have $\int_I f_k = \int_J \int_K f_k(x,y)dxdy$.

The sequence $\{f_k\}$ increases pointwise to f on I so both sequences

$$\left\{\int_I f_k\right\} \text{ and } \left\{\int_J f_k(x,y)dx\right\}, y \in K\,,$$

are increasing with $\int_I f_k = \int_J \int_K f_k(x,y)dxdy \le \int_J \int_K g(x,y)dxdy$ and $\int_J f_k(x,y)dx \le \int_J g(x,y)dx$ for $y \in K$. The MCT implies that f is integrable over I with $\lim \int_I f_k = \int_I f$ and that $f(\cdot,y)$ is integrable over J with $\lim \int_J f_k(\cdot,y) = \int_J f(\cdot,y)$. Also, the sequence $\{\int_J f_k(\cdot,y)\}$ is increasing with $\int_K \int_J f_k(x,y)dxdy \le \int_K \int_J g(x,y)dxdy$ so another application of MCT gives $\lim \int_K \int_J f_k(x,y)dxdy = \int_K \int_J f(x,y)dxdy$. Hence, $\int_I f = \int_K \int_J f(x,y)dxdy$ as desired.

Remark 10. *If f is a continuous function on a closed unbounded interval I, then the hypothesis in Theorem 9 is satisfied. For let $\{K_j\}$ be a sequence of compact subintervals of I such that $K_j \subset K_{j+1}$ and $I = \bigcup_{j=1}^\infty K_j$. First assume that f is non-negative. By Lemma 6 for every j there is a step function s_j vanishing outside K_j such that $|s_j(z) - f(z)| < 1/j$ for $z \in K_j$. Then $s_j \to f$ pointwise on I. For general f we can decompose $f = f^+ - f^-$ and apply the observation above to both f^+ and f^-.*

In applying Theorem 9 a good candidate for the function g is $|f|$. In particular, if $\int_K \int_J |f(x,y)|dxdy$ exists, then Theorem 9 implies that both f and $|f|$ are integrable over I and the conclusion of Theorem 9 holds for both f and $|f|$. Exercise 2 shows that it is important that the function g in Theorem 9 be non-negative.

As an application of Theorem 9 we show the existence of the convolution product of 2 continuous, absolutely integrable functions. Let $f, g : \mathbb{R} \to \mathbb{R}$ be continuous and absolutely integrable over $\mathbb{R}$. The *convolution* product of f and g is defined to be

$$f * g(x) = \int_{\mathbb{R}} f(x-y)g(y)dy$$

when the integral exists. We show $f * g$ exists for all $x \in \mathbb{R}$. Indeed, we have

$$\begin{aligned}\int_{\mathbb{R}}\int_{\mathbb{R}} |f(x-y)||g(y)|dxdy &= \int_{\mathbb{R}} |g(y)| \int_{\mathbb{R}} |f(x)|dxdy \\ &= \int_{\mathbb{R}} |g| \int_{\mathbb{R}} |f|,\end{aligned}$$

and by Theorem 9 the function $(x,y) \to f(x-y)g(y)$ is absolutely integrable over $\mathbb{R}^2$,

$$x \to \int_{\mathbb{R}} f(x-y)g(y)dy = f * g(x)$$

exists for all x and $f * g$ is integrable over $\mathbb{R}$. Moreover, the computation above shows $\int_{\mathbb{R}} |f * g| \le \int_{\mathbb{R}} |f| \int_{\mathbb{R}} |g|$.

FUBINI THEOREM*: GENERAL CASE

In this section we establish a very general form of the Fubini Theorem for the gauge integral which does not require the restrictive assumptions in Theorems 8 and 9. The proof does, however, use the property of null sets and integrable functions established in Corollary 6.6.

We first establish a lemma required for the existence of an iterated integral.

Lemma 11. *Let $f : I \to \mathbb{R}$ be integrable over I and let*

$$N = \{y \in K : f(\cdot, y) \text{ is not integrable over } J\}.$$

Then N is null.

Proof: $y \in N$ if and only if there exists $\varepsilon > 0$ such that for every gauge γ_J on J there exist γ_J-fine tagged partitions $\mathcal{P}, \mathcal{Q}$ of J such that

(*) $S(f(\cdot,y),\mathcal{P}) - S(f(\cdot,y),\mathcal{Q}) \geq \varepsilon.$

For each $i \in \mathbb{N}$ let N_i be the set of all $y \in N$ which satisfy condition $(*)$ with $\varepsilon = 1/i$. Then $N_i \subset N_{i+1}$ and $N = \bigcup_{i=1}^{\infty} N_i$ so it suffices to show that each N_i is null (Exercise 1.9).

Fix i and let $\varepsilon > 0$. To show N_i is null it suffices to show there exists a gauge γ_K on K such that $|S(C_{N_i},\mathcal{R})| \leq \varepsilon$ whenever $\mathcal{R}$ is a γ_K-fine tagged partition of J, i.e., $\int_K C_{N_i} = 0$ [6.6].

There exists a gauge γ on I such that

(**) $|S(f,\mathcal{D}) - S(f,\mathcal{E})| < \varepsilon/i$

whenever $\mathcal{D}$ and $\mathcal{E}$ are γ-fine tagged partitions of I. If $y \in N_i$, $\gamma_y(x) = P_1\gamma(x,y)$ defines a gauge γ_y on J. By definition of N_i there exist 2 γ_y-fine tagged partitions $\mathcal{P}_y$ and $\mathcal{Q}_y$ of J such that

(***) $S(f(\cdot,y),\mathcal{P}_y) - S(f(\cdot,y),\mathcal{Q}_y) \geq 1/i.$

If $y \in K \setminus N_i$, $\gamma_y(x) = P_1\gamma(x,y)$ still defines a gauge on J so pick a γ_y-fine tagged partition $\mathcal{P}_y$ of J and set $\mathcal{P}_y = \mathcal{Q}_y$ [note $S(f(\cdot,y),\mathcal{P}_y) - S(f(\cdot,y),\mathcal{Q}_y) = 0$ for $y \in K \setminus N_i$].

If $\mathcal{P}_y = \{(p_j^y, P_j^y) : 1 \leq j \leq p_y\}$ and $\mathcal{Q}_y = \{(q_j^y, Q_j^y) : 1 \leq j \leq q_y\}$, set

$$\gamma_2(y) = \left(\bigcap_{j=1}^{p_y} P_2\gamma(p_j^y, y)\right) \cap \left(\bigcap_{j=1}^{q_y} P_2\gamma(q_j^y, y)\right).$$

Then γ_2 is a gauge on K.

Suppose that $\mathcal{R} = \{(y_k, K_k) : 1 \leq k \leq m\}$ is a γ_2-fine tagged partition of K. By construction

$$\mathcal{D} = \{((p_j^{y_k}, y_k), P_j^{y_k} \times K_k) : 1 \leq j \leq p_{y_k}, 1 \leq k \leq m\}$$

and

$$\mathcal{E} = \{((q_j^{y_k}, y_k), Q_j^{y_k} \times K_k) : 1 \le j \le q_{y_k}, 1 \le k \le m\}$$

are both γ-fine.

Hence, by (**) and (***),

$$\begin{aligned}
\varepsilon/i &\ge |S(f,\mathcal{D}) - S(f,\mathcal{E})| \\
&= \left| \sum_{\substack{k=1 \\ y_k \in N_i}}^{m} \ell(K_k) \left\{ \sum_{j=1}^{p_{y_k}} f(p_j^{y_k}, y_k)\ell(P_j^{y_k}) - \sum_{j=1}^{q_{y_k}} f(q_j^{y_k}, y_k)\ell(Q_j^{y_k}) \right\} \right| \\
&= \left| \sum_{\substack{k=1 \\ y_k \subset N_i}}^{m} \ell(K_k)\{S(f(\cdot, y_k), \mathcal{P}_{y_k}) - S(f(\cdot, y_k), \mathcal{Q}_{y_k})\} \right| \\
&\ge 1/i \sum_{\substack{k=1 \\ y_k \in N_i}}^{m} \ell(K_k) = 1/i \sum_{k=1}^{m} C_{N_i}(y_k)\ell(K_k) = 1/i \; S(C_{N_i}, \mathcal{R})
\end{aligned}$$

so $S(C_{N_i}, \mathcal{R}) \le \varepsilon$ as desired.

Exercise 22 shows that the conclusion of Lemma 11 is best possible in some sense.

Let the notation be as in Lemma 11 and set $F(y) = \int_J f(x,y)dx$ for $y \notin N$. If we are to obtain a Fubini Theorem in the form of Theorem 8 or 9 we must extend the function F to all of K in such a way that the extended function is integrable over K with integral equal to $\int_I f$. Since N is a null set, if such an integrable extension exists it is independent of the way in which the extension is defined on K (Exercise 2.22). In the interest of simplicity, one usually agrees to define $F(y) = 0$ for $y \in N$. The resulting function, F, is then shown to be integrable over K with integral equal to $\int_I f$, and one writes

$$\int_I f = \int_K F(y)dy = \int_K \int_J f(x,y)dxdy\,.$$

The following technical lemma is useful in the proof of the general Fubini Theorem.

Lemma 12. *If $N \subset \mathbb{R}$ is null, then $N \times \mathbb{R}$ is null (in $\mathbb{R}^2$).*

Proof: It suffices to show that $N_k = N \times (-k, k)$ is null for every $k \in \mathbb{N}$ since

$$N \times \mathbb{R} = \cup\{N_k : k \in N\}$$

(Exercise 1.9). Fix k and let $\varepsilon > 0$. There exists a sequence of open intervals $\{J_j\}$ in $\mathbb{R}$ such that $N \subset \bigcup_{j=1}^{\infty} J_j$ and $\sum_{j=1}^{\infty} \ell(J_j) < \varepsilon$. Then $\{I_j\} = \{J_j \times (-k, k)\}$ is a sequence of open intervals in $\mathbb{R}^2$ such that $N_k \subset \bigcup_{j=1}^{\infty} I_j$ and $\sum_{j=1}^{\infty} v(I_j) < \varepsilon k$. Hence, N_k is null.

We now state and prove a general form of Fubini's Theorem for the gauge integral; the proof uses the construction of compound tagged partitions given in Lemma 1.

Theorem 13 (*Fubini*). *Let $f : I \to \mathbb{R}$ be integrable over I and $F(y) = \int_J f(x, y)dx$, for $y \in K$, be defined as above. Then F is integrable over K with*

$$\int_I f = \int_K F(y)dy = \int_K \int_J f(x, y)dxdy\,.$$

Proof: From Lemmas 11 and 12 there is a function $g : I \to \mathbb{R}$ such that $f = g$ a.e. in I, $f(\cdot, y) = g(\cdot, y)$ whenever $f(\cdot, y)$ is integrable over J, $g(\cdot, y)$ is integrable over J for every $y \in K$ and $F(y) = \int_J g(x, y)dx$ for $y \in K$ (Exercise 2.22). Thus, we may assume that $f(\cdot, y)$ is integrable for every $y \in K$.

Let $\varepsilon > 0$ and choose a gauge γ on I such that $|S(f, \mathcal{D}) - \int_I f| < \varepsilon$ whenever $\mathcal{D}$ is a γ-fine tagged partition of I. By Lemma 5.3 pick a function $\varphi : K \to (0, \infty)$ which is integrable on K and a gauge γ_φ on K such that $0 \leq S(\varphi, \mathcal{P}) \leq 1$ whenever $\mathcal{P}$ is a γ_φ-fine tagged partition of K. For each $y \in K$ pick a $\gamma_1(\cdot, y)$-fine tagged partition $\mathcal{P}_y$ of J such that

$$\left| S(f(\cdot, y), \mathcal{P}_y) - \int_J f(x, y)dx \right| < \varepsilon\varphi(y)$$

[integrability of $f(\cdot, y)$]. Put

$$h(y) = \int_J f(x, y)dx - S(f(\cdot, y), \mathcal{P}_y)\,.$$

Define $\gamma_2'(y) = \cap\{\gamma(x, y) : (x, P) \in \mathcal{P}_y\}$ as in Lemma 1, and set $\gamma_2^* = \gamma_\varphi \cap \gamma_2'$. We show that $|S(F, \mathcal{K}) - \int_I f| < \varepsilon$ whenever $\mathcal{K}$ is a γ_2^*-fine tagged partition of K, and this will establish the result. Let $\mathcal{D}$ be the γ-fine compound tagged partition generated by $\mathcal{K}$ and $\{\mathcal{P}_y : y \in K\}$ as in Lemma 1. Then

$$\begin{aligned}\left|S(F,\mathcal{K}) - \int_I f\right| &\le |S(F,\mathcal{K}) - S(f,\mathcal{D})| + \left|S(f,\mathcal{D}) - \int_I f\right| \\ &< \left|\sum_{(y,K)\in\mathcal{K}} \{F(y)\ell(K) - S(f(\cdot,y),\mathcal{P}_y)\ell(K)\}\right| + \varepsilon \\ &= \left|\sum_{(y,K)\in\mathcal{K}} h(y)\ell(K)\right| + \varepsilon \le \sum_{(y,K)\in\mathcal{K}} \varepsilon\varphi(y)\ell(K) + \varepsilon \\ &= \varepsilon S(\varphi,\mathcal{K}) + \varepsilon \le 2\varepsilon\,.\end{aligned}$$

Theorem 13 has an interesting application to the computation of Lebesgue measure in $\mathbb{R}^2$. [It is straightforward to extend the definition and terminology of (Lebesgue) integrable and measurable subsets of $\mathbb{R}^2$ as was done in Chapter 6.] As in Chapter 6 let λ be Lebesgue measure on $\mathbb{R}$, and let λ_2 be Lebesgue measure on $\mathbb{R}^2$.

If $E \subset \mathbb{R}^2$ and $y \in \mathbb{R}$, the *y-section* of E at y is defined to be

$$E^y = \{x : (x,y) \in E\}\,.$$

Note $C_E(x,y) = C_{E^y}(x)$ for $x, y \in \mathbb{R}$. From Lemma 11 and Theorem 13 we obtain:

Corollary 14. *Let $E \subset \mathbb{R}^2$ be integrable. Then E^y is integrable over $\mathbb{R}$ for all y, except those in a null set. With the same notation as in Theorem 13 the function $y \to \lambda(E^y)$ is integrable over $\mathbb{R}$ with $\lambda_2(E) = \int_{\mathbb{R}} \lambda(E^y)dy$.*

Notes/Remarks

The Fubini Theorem along with the convergence theorems (MCT, DCT) are the great successes of the Lebesgue integration theory. The versions of the Fubini/Tonelli Theorems given in Theorems 8 and 9 although applicable to a much smaller class of functions than in the Lebesgue case, are sufficient for an introductory real analysis course since most functions encountered are continuous (or "nearly continuous") (see [Sw1] 3.9.3, 3.9.4 for the Lebesgue case). The general form of the Fubini Theorem for the gauge integral given in Theorem 13 illustrates the generality of the gauge integral since the theorem applies to conditionally convergent integrals (Exercises 6 and 25) unlike the absolutely convergent Lebesgue integral.

The Fubini Theorem gives a method for computing a "double integral" as an iterated integral assuming the function is integrable. A Tonelli Theorem is a theorem which establishes the integrability of a function from assumptions about the existence of iterated integrals; Theorem 9 is such a "Tonelli result." We establish a more general form of the Tonelli Theorem for measurable functions.

Theorem 15* (*Tonelli*). *Let $f : I \to \mathbb{R}$ be measurable. If $\int_K \int_J |f(x,y)|dxdy$ exists and is finite, then f is integrable over I with $\int_I f = \int_K \int_J f(x,y)dxdy$.*

Proof: By decomposing $f = f^+ - f^-$, we may assume that f is non-negative.

For each k set $I_k = \{(x,y) : |x| \leq k, |y| \leq k\} \cap I$. There exists a sequence $\{\varphi_k\}$ of non-negative, simple functions such that $\varphi_k \uparrow f$. ([Sw1] 3.1.1.2.) Set $f_k = C_{I_k}\varphi_k$. Then f_k is non-negative, simple, integrable and $f_k \uparrow f$. By Corollary 14 and the linearity of the integral, the conclusion of Fubini's Theorem 8 holds for f_k. The proof of the theorem now follows as the proof of Theorem 9.

For further examples and counter-examples relative to Fubini's Theorem, see [Sw1] §3.9.

As noted earlier it is straightforward to extend the development of Lebesgue measure in $\mathbb{R}$ that was carried out in Chapter 6 to $\mathbb{R}^n$. For more details on Lebesgue measure in $\mathbb{R}^n$, see [Sw1], [N2].

The convolution product of 2 functions as defined following Remark 10 is applicable to a much wider class of functions than considered there. For details on the existence of convolution products, see [Sw1] 3.11.1, 6.1.23, [HS]; for applications of the convolution product, see [Sw1] §3.11, [DS] §18, [HS].

Exercises

1. Let J, K be compact intervals in $\mathbb{R}$ and $I = J \times K$. Let γ be a gauge on I. Use the following argument to show that there exists a γ-fine tagged partition of I. [Suppose the result false and divide I into 4 "equal" subintervals. The result must be false for one of the subintervals. Now continue to divide. In $\mathbb{R}^n$ the same construction works by dividing into 2^n "equal" subintervals.]
2. Let $f(x,y) = e^{-xy} - 2e^{-2xy}$. Show that $\int_0^1 \int_1^\infty f(x,y)dxdy > 0$ and $\int_1^\infty \int_0^1 f(x,y)dydx < 0$.

3. Let $f(x,y) = xy/(x^2+y^2)^2$ when $(x,y) \neq (0,0)$ and $f(0,0) = 0$. Show that

$$\int_{-1}^{1}\int_{-1}^{1} f(x,y)dxdy = \int_{-1}^{1}\int_{-1}^{1} f(x,y)dydx = 0$$

but f is not integrable over $[-1,1] \times [-1,1]$. [Hint: Show that f is not integrable over $[0,1] \times [0,1]$ by showing

$$\int_{1/n}^{1}\int_{0}^{1} f(x,y)dxdy = -1/4 \ln\, 2 + 1/4 \ln\, (1+1/n^2) - 1/2\, \ln\, 1/n$$

$$= \int_{E_n} f\,,$$

where $E_n = \{(x,y) : 0 \leq x \leq 1, 1/n \leq y \leq 1\}$, and $\{\int_{E_n} f\}$ is unbounded.]
4. Let J, K be closed intervals in $\mathbb{R}^*$ and $I = J \times K$. Let $\mathcal{J}, \mathcal{K}$ be tagged partitions of J, K, respectively. Show $\mathcal{I} = \{((x,y), A \times B) : (x,A) \in \mathcal{J}, (y,B) \in \mathcal{K}\}$ is a tagged partition of I. Give an example of a tagged partition of I which is not of this form. Let $\gamma_J(\gamma_K)$ be a gauge on $J(K)$. Show that $\gamma(x,y) = \gamma(x) \times \gamma(y)$ is a gauge on I, and if $\mathcal{J} << \gamma_J$, $\mathcal{K} << \gamma_K$, then $\mathcal{I} << \gamma$. Give an example of a gauge on I which is not of this form.
5. Use only the definition of the integral to show that $C_{[0,1]}$ is integrable over $\mathbb{R}^2$.
6. Show that the function f in Example 5 is absolutely integrable if and only if the series $\sum a_k$ is absolutely convergent.
7. Extend the definition of a null set to $\mathbb{R}^2(\mathbb{R}^n)$. If N is a null subset of $\mathbb{R}^2$, use only the definition of the integral to show that C_N is integrable over $\mathbb{R}^2$.
8. Suppose f satisfies the hypothesis of Lemma 6. Show that there exists a sequence of non-negative step functions $\{s_k\}$ such that $s_k(x) \uparrow f(x)$ for $x \in I$, where the convergence is uniform on K.
9. Let $\alpha, \beta : [a,b] \to \mathbb{R}$ be continuous with $\alpha(t) \leq \beta(t)$ for $a \leq t \leq b$. Let

$$E = \{(x,y) : a \leq x \leq b,\ \alpha(x) \leq y \leq \beta(x)\}\,.$$

Suppose $f : \mathbb{R}^2 \to \mathbb{R}$ is continuous. Show that f is integrable over E and $\int_E f = \int_a^b \int_{\alpha(x)}^{\beta(x)} f(x,y)dydx$.
10. Find the integral of the function $f(x,y) = x^3y^2$ over the region in the first quadrant inside the ellipse $9x^2 + 25y^2 = 25 \cdot 9$ and outside the circle $x^2 + y^2 = 1$.

11. Define $f : [0,1] \times [0,1] \to \mathbb{R}$ by $f(x,y) = 1$ if $x, y \in \mathbb{Q}$ and $f(x,y) = 0$ otherwise. Verify Theorem 13 for f. Do any of the integrals exist as Riemann integrals? What can you conclude?
12. Show that the function $f(x,y) = e^{-x} \sin y$ is integrable over $[0,\infty] \times [0, 2\pi]$ and calculate the integral.
13. Is the function $f(x,y) = 1/(x+y)$ integrable over $[0,1] \times [0,1]$? What about $f(x,y) = 1/(x^2+y^2)$?
14. Give an example of a function f such that $\int_0^1 \int_0^1 f(x,y)dxdy$ exists but $\int_0^1 \int_0^1 f(x,y)dydx$ does not.
15. Let $f : J \to \mathbb{R}, g : K \to \mathbb{R}$ and define $h : I = J \times K \to \mathbb{R}$ by $h(x,y) = f(x)g(y)$. Assume both f and g are pointwise limits of step functions and both are absolutely integrable. Show that h is absolutely integrable with $\int_I h = \int_J f \int_K g$.
16. Evaluate $\int_0^1 \int_{3y}^3 e^{x^2} dxdy$. [Hint: Change the order of integration.]
17. Let $E_n = [0,n] \times [0,n]$. Show that $\int_0^\infty \frac{\sin x}{x} dx = \pi/2$ by evaluating $\lim \int_{E_n} e^{-xy} \sin x$ in 2 different ways.
18. Let $f(x,y) = ye^{-(1+x^2)y^2}$. Show $\int_0^\infty \int_0^\infty f(x,y)dxdy = \int_0^\infty \int_0^\infty f(x,y)dydx$ and use this to show that $\int_0^\infty e^{-x^2} dx = \sqrt{\pi}/2$.
19. Let $f(x,y) = e^{-xy}/(1+x^2)$. Show that f is integrable over $[0,\infty] \times [0,1]$.
20. Let $f(x,y) = e^{-xy} \sin xy$. Show that f is integrable over $[0,\infty] \times [1,2]$.
21. Extend the first part of Corollary 14 to measurable subsets.
22. Assume the existence of a subset $P \subset [0,1]$ which is not Lebesgue measurable ([Sw1] 1.3.1). Let $Z \subset [0,1]$ be null and set $E = Z \times P$. Show that C_E is integrable over $[0,1] \times [0,1]$ but $C_E(\cdot, y)$ is not integrable over $[0,1]$ for all y. [See Lemma 11.]
23. Let P be as in Exercise 22. Show that $[0,1] \times P$ is not integrable (measurable) in $\mathbb{R}^2$.
24. Analogous to Exercise 1.12 show that when $n = 2$ it can be assumed in Definition 8.3 that all tags are vertices of the subintervals which they tag. What about general n?
25. Verify the conclusion of the Fubini Theorem 13 for the function in Example 5.
26. (Improper Integral) Let $I_k = [0, 1/2^k]$, $k = 0, 1, \ldots$, and $J_k = I_k \times I_k$. Suppose that $f : J_0 \to \mathbb{R}$ is integrable over $J_0 \setminus J_k$ for $k \geq 1$ and $\lim \int_{J_0 \setminus J_k} f = L$ exists. Show that f is integrable over J_0 with $\int_{J_0} f = L$. (Compare with Theorem 3.4.)
27. Apply Exercises 26 to Example 5.
28. (Linear Change of Variable) Let $f : \mathbb{R}^n \to \mathbb{R}$ be absolutely integrable and

$A : \mathbb{R}^n \to \mathbb{R}^n$ be linear. Show that $f \circ A$ is absolutely integrable with $\int_{\mathbb{R}^n} f \circ A |\det A| = \int_{\mathbb{R}^n} f$. [Hint: A is a product of transformations of the form $(x_1, \dots, x_n) \to (tx_1, \dots, x_n), (x_1, \dots, x_n) \to (x_1 + x_2, x_2, \dots, x_n)$, $(x_1, \dots, x_i, \dots, x_j, \dots, x_n) \to (x_1, \dots, x_j, \dots, x_i, \dots, x_n)$. Use Fubini's Theorem.]

29. Show that absolute integrability is important in Exercise 27. [Hint: Let $I_k = [\frac{1}{2^{k-1}}, \frac{1}{2^k}]$ for $k \in \mathbb{N}$ and $J_k = I_k \times I_k$. Let $A_k = \{(t_1, t_2) \in J_k : t_2 > t_1\}$, $B_k = \{(t_1, t_2) \in J_k : t_2 < t_1\}$. Construct a function f which is 0 outside $\bigcup_{k=0}^{\infty} J_k$ and is integrable over each J_k with $\int_{A_k} f = -\int_{B_k} f = 1/k$. Use Exercise 26 to show f is integrable over $\mathbb{R}^2$. Define $A : \mathbb{R}^2 \to \mathbb{R}^2$ by $A(t_1, t_2) = (\frac{t_1 - t_2}{\sqrt{2}}, \frac{t_1 + t_2}{\sqrt{2}})$ [rotation through $\pi/4$]. Show that $f \circ A$ is not integrable.]

Chapter 9

The McShane Integral

9.1 Definition and Basic Properties

In this chapter we describe another gauge-type integral due to E.J. McShane, which we will call the McShane integral. McShane alters the definition of the gauge or Henstock-Kurzweil integral by not requiring that the tag associated with a subinterval in a tagged partition belong to the associated subinterval, that is, the tag is free to belong anywhere in the domain of the function being considered. This requirement is somewhat non-intuitive; it does not seem to be known what led McShane to this definition, but it does have a profound effect on the subsequent integration theory which is developed. In fact, the McShane integral is equivalent to the Lebesgue integral in Euclidean spaces [Appendix 4].

We will begin by defining the McShane integral over closed subintervals in $\mathbb{R}^*$. Let I be a closed interval (bounded or unbounded) in $\mathbb{R}^*$. Again if f is a real-valued function, it is always assumed that f is extended to all of $\mathbb{R}^*$ with $f(\pm\infty) = 0$. A *partition* of I is a finite collection of closed, non-overlapping subintervals of I whose union is I. A *free tagged partition* of I is a finite collection of pairs $\mathcal{D} = \{(t_i, I_i) : 1 \le i \le m\}$ such that $\{I_i : 1 \le i \le m\}$ is a partition of I and $t_i \in I$; we again refer to the point t_i as being the *tag* associated with the subinterval I_i. We refer to such tagged partitions as free tagged partitions to distinguish them from the tagged partitions used in the gauge or Henstock-Kurzweil integral. If $f : I \to \mathbb{R}$ and $\mathcal{D} = \{(t_i, I_i) : 1 \le i \le m\}$ is a free tagged partition of I, we define the *Riemann sum* of f with respect to $\mathcal{D}$ as before to be $S(f, \mathcal{D}) = \sum_{i=1}^{m} f(t_i)\ell(I_i)$.

If γ is a gauge on I, we say that a free tagged partition $\mathcal{D} = \{(t_i, I_i) : 1 \le i \le m\}$ is *γ-fine* if $I_i \subset \gamma(t_i)$ for $1 \le i \le m$, and as before we write $\mathcal{D} << \gamma$ when $\mathcal{D}$ is γ-fine.

Definition 1. *A function $f : I \to \mathbb{R}$ is McShane integrable over I if there exists $A \in \mathbb{R}$ such that for every $\varepsilon > 0$ there is a gauge γ on I such that $|S(f, \mathcal{D}) - A| < \varepsilon$ whenever $\mathcal{D}$ is a γ-fine free tagged partition of I.*

From Theorem 1.3 or Theorem 4.1 it follows that the definition of the McShane integral makes sense [i.e., any gauge has a γ-fine (free) tagged partition], and as was the case for the gauge integral, the number A is unique and is called the McShane integral of f over I. In this chapter we will only consider the McShane integral so we will write $A = \int_I f$, etc., as before. In situations where other integrals are considered we will adopt notation which will make it clear which integral is being considered [e.g., Chapter 10 and Appendix 4]. Henceforth, we will refer to the Henstock-Kurzweil integral as the gauge integral to distinguish it from the McShane integral.

Since there are more free tagged partitions than (gauge) tagged partitions, it is clear that any McShane integrable function is gauge integrable and the two integrals are equal. We will see below there are functions which are gauge integrable but not McShane integrable so the use of free tagged partitions does lead to a less general integration theory; in particular, we will see that the use of free tagged partitions implies that the McShane integral is an "absolute integral" [Theorem 5] in contrast to the "conditional" gauge integral (recall Examples 2.12 and 4.8).

There are a couple of observations concerning the use of free tagged partitions which should be made. First, one can no longer assume that each tag only appears at most twice or that the tag appears only as an endpoint (Exercise 1.12) as was the case for (gauge) tagged partitions (see Exercise 1). Although the tag is not required to belong to its associated subinterval in a free tagged partition, a gauge γ does somehow require the tag to be "near" its associated subinterval.

The proofs of many of the basic properties of the McShane integral are essentially identical to the proofs of the corresponding properties of the gauge integral. We will list these basic properties in Theorem 2 below; the reader is invited to check the corresponding proofs for the gauge integral.

Theorem 2. *Let $f, g : I \to \mathbb{R}$ be McShane integrable over I.*

(i) *$f + g$ is McShane integrable over I with $\int_I (f + g) = \int_I f + \int_I g$.*
(ii) *For every $a \in \mathbb{R} af$ is McShane integrable over I with $\int_I af = a \int_I f$.*
(iii) *If $f \geq g$ on I, then $\int_I f \geq \int_I g$.*

For the proof, see Theorem 2.1.

Theorem 3. *Let $f : I \to \mathbb{R}$.*

(i) (*Cauchy Criterion*) *f is McShane integrable over I if and only if for every $\varepsilon > 0$ there is a gauge γ on I such that $|S(f, \mathcal{D}_1) - S(f, \mathcal{D}_2)| < \varepsilon$ whenever $\mathcal{D}_1$ and $\mathcal{D}_2$ are γ-fine free tagged partitions of I.*

(ii) *If f is McShane integrable over I and J is a closed subinterval of I, then f is McShane integrable over J.*

(iii) *If $I = [a, b]$ and $-\infty \le a < c < b \le \infty$ and f is McShane integrable over I, then f is McShane integrable over $[a, c]$ and $[c, b]$ with $\int_a^b f = \int_a^c f + \int_c^b f$.*

(iv) *If for every $\varepsilon > 0$ there exist McShane integrable functions $g, h : I \to \mathbb{R}$ with $g \le f \le h$ and $\int_I (h - g) < \varepsilon$, then f is McShane integrable over I.*

(v) *If I is a closed, bounded interval and f is continuous on I, then f is McShane integrable over I.*

(vi) (*Henstock's Lemma*). *Suppose f is McShane integrable over I and for $\varepsilon > 0 \gamma$ is a gauge on I such that $|S(f, \mathcal{D}) - \int_I f| < \varepsilon$ whenever $\mathcal{D}$ is a γ-fine free tagged partition of I. If $\mathcal{J} = \{(t_i, J_i) : 1 \le i \le m\}$ is any γ-fine free partial tagged partition of I and if $J = \bigcup_{i=1}^m J_i$, then*

$$\left| S(f, \mathcal{J}) - \int_J f \right| \le \varepsilon \text{ and } \sum_{i=1}^m \left| f(t_i)\ell(J_i) - \int_{J_i} f \right| \le 2\varepsilon .$$

[*Free partial tagged partitions, Riemann sums for free partial tagged partitions and γ-fine free partial tagged partitions are defined as in Chapter* 3.]

For the proofs of (i)–(vi), see, respectively, Theorems 2.6, 2.7, 2.4, Lemma 2.9, Theorem 2.10 and Lemma 3.1.

We now come to the basic difference in the McShane and gauge integrals; the McShane integral is an "absolute" integral. For the proof of this result we need a preliminary lemma; the reader should take note of the use of free tagged partitions in the proof.

Lemma 4. *Let $f : I \to \mathbb{R}$ be McShane integrable over I and let $\varepsilon > 0$. Suppose γ is a gauge on I such that for every γ-fine free tagged partition $\mathcal{D}$ of I, we have $|S(f, \mathcal{D}) - \int_I f| < \varepsilon$. If $\mathcal{D} = \{(t_i, I_i) : 1 \le i \le m\}$ and $\mathcal{E} = \{(s_j, J_j) : 1 \le j \le n\}$ are γ-fine free tagged partitions of I, then*

$$\sum_{i=1}^m \sum_{j=1}^n |f(t_i) - f(s_j)| \ell(I_i \cap J_j) \le 2\varepsilon .$$

Proof: Put $\mathcal{F} = \{I_i \cap J_j = K_{ij} : 1 \le i \le m, 1 \le j \le n, I_i^\circ \cap J_j^\circ \ne \emptyset\}$. Define 2 γ-fine free tagged partitions $\mathcal{D}'$ and $\mathcal{E}'$ of I as follows: if $f(t_i) \ge f(s_j)$, put $t_{ij} = t_i$ and $s_{ij} = s_j$; if $f(t_i) < f(s_j)$, put $t_{ij} = s_j$ and $s_{ij} = t_i$. Note $f(t_{ij}) - f(s_{ij}) = |f(t_i) - f(s_j)|$. Now set $\mathcal{D}' = \{(t_{ij}, K_{ij}) : K_{ij} \in \mathcal{F}\}$ and $\mathcal{E}' = \{(s_{ij}, K_{ij}) : K_{ij} \in \mathcal{F}\}$. Note $\mathcal{D}'$ and $\mathcal{E}'$ are γ-fine so by hypothesis

$$|S(f, \mathcal{D}') - S(f, \mathcal{E}')| = \left| \sum_{K_{ij} \in \mathcal{F}} \{f(t_{ij}) - f(s_{ij})\} \ell(K_{ij}) \right|$$

$$= \sum_{i=1}^{m} \sum_{j=1}^{n} |f(t_i) - f(s_j)| \ell(I_i \cap J_j) \le 2\varepsilon \,.$$

Theorem 5. *If $f : I \to \mathbb{R}$ is McShane integrable over I, then $|f|$ is McShane integrable over I with $|\int_I f| \le \int_I |f|$.*

Proof: We show that the Cauchy criterion is satisfied for $|f|$. Let $\varepsilon > 0$. Let γ be a gauge on I such that $|S(f, \mathcal{D}) - \int_I f| < \varepsilon$ whenever $\mathcal{D}$ is a γ-fine free tagged partition of I. Let $\mathcal{D} = \{(t_i, I_i) : 1 \le i \le m\}$ and $\mathcal{E} = \{(s_j, J_j) : 1 \le j \le n\}$ be γ-fine free tagged partitions of I. Then by Exercise 1 and Lemma 4.

(*)

$$|S(|f|, \mathcal{D}) - S(|f|, \mathcal{E})| = \left| \sum_{i=1}^{m} \sum_{j=1}^{n} \{|f(t_i)| - |f(s_j)|\} \ell(I_i \cap J_j) \right|$$

$$\le \sum_{i=1}^{m} \sum_{j=1}^{n} |f(t_i) - f(s_j)| \ell(I_i \cap J_j) \le 2\varepsilon$$

so the Cauchy criterion is satisfied.

Since $f \le |f|$ and $-f \le |f|$, the last inequality follows directly from Theorem 2 (iii).

Theorem 5 should be contrasted with Theorem 4.10 for the gauge integral; recall Examples 2.12, 4.8 and Exercises 3.9, 4.3.

Corollary 6. *Let $f, g : I \to \mathbb{R}$ be McShane integrable over I. Then $f \wedge g$ and $f \vee g$ are McShane integrable over I.*

Proof: $f \vee g = (f + g + |f - g|)/2$ and $f \wedge g = -((-f) \vee (-g))$.

This result should also be compared with Proposition 4.12; recall Examples 2.12 and 4.8.

The proof of Theorem 5 actually allows us to establish a slightly more general result which has some interesting corollaries. A function g defined on an interval J satisfies a *Lipschitz condition* if there exists $L > 0$ such that $|g(t) - g(s)| \leq L|s - t|$ for all $s, t \in J$; the parameter L is called a *Lipschitz constant* for g. [See Exercise 6 for examples of functions satisfying Lipschitz conditions.]

Theorem 7. *Let $f : I \to \mathbb{R}$ be McShane integrable over I and let $g : J \to \mathbb{R}$ satisfy a Lipschitz condition over the interval J. Assume $J \supset \{f(t) : t \in I\}$. Then $g \circ f$ is integrable over I.*

Proof: If we replace f with $g \circ f$ in the inequality (*), we obtain

$$|S(g \circ f, \mathcal{D}) - S(g \circ f, \mathcal{E})| \leq L2\varepsilon\,,$$

where L is a Lipschitz constant for g. Therefore, $g \circ f$ is integrable by the Cauchy criterion.

See Exercise 9 for the analogue of the inequality in Theorem 5.

Corollary 8. *Let $f : I \to \mathbb{R}$ be McShane integrable over I and $c > 0$. Then $c \wedge f$ and $(-c) \vee f$ are McShane integrable over I.*

Proof: Let $g(t) = c \wedge t$ for $t \in \mathbb{R}$. Then g satisfies a Lipschitz condition on $\mathbb{R}$ (Exercise 8). Thus, Theorem 7 is applicable and gives that $c \wedge f$ is integrable over I. The other part is similar using the function $t \to (-c) \vee t$.

We next consider the Fundamental Theorem of Calculus (FTC) for the McShane integral. First, we need a result which is just a rephrasing of the FTC for the gauge integral (Theorem 1.5).

Lemma 9. *Let $f : [a, b] \to \mathbb{R}$ be differentiable on $[a, b]$ and $\varepsilon > 0$. There exists a gauge γ on $[a, b]$ such that $|S(f', \mathcal{D}) - (f(b) - f(a))| < \varepsilon$ whenever $\mathcal{D}$ is a γ-fine (gauge) tagged partition of $[a, b]$.*

Theorem 10 (*FTC*; *Part* 1). *Let $f : [a, b] \to \mathbb{R}$ be differentiable on $[a, b]$ and assume that f' is McShane integrable over $[a, b]$. Then $\int_a^b f' = f(b) - f(a)$.*

Proof: Let $\varepsilon > 0$. There exists a gauge γ_1 on $[a, b]$ such that

$$\left| S(f', \mathcal{D}) - \int_a^b f' \right| < \varepsilon$$

whenever $\mathcal{D}$ is a γ_1-fine free tagged partition of $[a, b]$. Let γ be the gauge in Lemma 9. Put $\gamma_2 = \gamma_1 \cap \gamma$ and let $\mathcal{D}$ be a γ-fine (gauge) tagged partition of

$[a, b]$. From Lemma 9, we have

$$\left|\int_a^b f' - (f(b) - f(a))\right| \le \left|\int_a^b f' - S(f', \mathcal{D})\right| + |S(f', \mathcal{D}) - (f(b) - f(a))| < 2\varepsilon$$

and the result follows.

This version of the FTC should be contrasted with the FTC for the gauge integral (Theorem 1.5) where it is not necessary to assume the integrability of the derivative. Example 2.12 along with Theorem 5 shows that the assumption of the McShane integrability of the derivative in Theorem 10 cannot be dropped.

For part 2 of the FTC we have the analogue of Theorem 2.8.

Theorem 11 (*FTC; Part 2*). *Let $f : [a, b] \to \mathbb{R}$ be McShane integrable over $[a, b]$ and set $F(t) = \int_a^t f$. If f is continuous at $t_0 \in [a, b]$, then F is differentiable at t_0 with $F'(t_0) = f(t_0)$.*

See the proof of Theorem 2.8.

We next consider an interesting approximation result which is also used later in a convergence theorem.

Theorem 12. *Let $f : I \to \mathbb{R}$ be McShane integrable over I and $\varepsilon > 0$. There exists a McShane integrable step function $g : I \to \mathbb{R}$ such that $\int_I |f - g| < \varepsilon$.*

Proof: There exists a gauge γ_1 on I with $\gamma_1(t)$ bounded for every $t \in \mathbb{R}$ such that $|S(f, \mathcal{D}) - \int_I f| < \varepsilon/3$ whenever $\mathcal{D}$ is a γ_1-fine free tagged partition of I. Fix such a partition $\mathcal{D} = \{(t_i, I_i) : 1 \le i \le m\}$. Define a step function $g : I \to \mathbb{R}$ by setting $g = \sum_{i=1}^m f(t_i) C_{I_i}$ [note $\pm\infty$ must be a tag for any unbounded interval and $f(\pm\infty) = 0$].

Now $|f - g|$ is McShane integrable over I (Theorems 2 and 5) so there exists a gauge γ_2 on I such that $|S(|f - g|, \mathcal{E}) - \int_I |f - g|| < \varepsilon/3$ whenever $\mathcal{E}$ is a γ_2-fine free tagged partition of I. Set $\gamma = \gamma_1 \cap \gamma_2$. In each subinterval I_k, let $\mathcal{E}_k$ be a γ-fine (gauge) tagged partition of I_k (Theorems 1.3 and 4.1). Set $\mathcal{E} = \bigcup_{k=1}^m \mathcal{E}_k$ and note that $\mathcal{E}$ is a γ-fine (gauge) tagged partition of I. Suppose that $\mathcal{E} = \{(s_k, J_k) : 1 \le k \le n\}$. For each $k, 1 \le k \le n$, there is a unique j_k such that $J_k \subset I_{j_k}$. Consider the pair (t_{j_k}, J_k): we have $s_k \in J_k \subset I_{j_k} \subset \gamma_1(t_{j_k})$ so $\mathcal{P} = \{(t_{j_k}, J_k) : 1 \le k \le n\}$ is γ_1-fine. Since $\mathcal{E}$ is also γ_1-fine, Lemma 4 gives

$$\sum_{k=1}^n \sum_{i=1}^n |f(s_i) - f(t_{j_k})| \ell(J_i \cap J_k) \le 2\varepsilon/3\,.$$

But, $\ell(J_i \cap J_k) = 0$ if $i \neq k$ so

$$\sum_{k=1}^{n} |f(s_k) - f(t_{j_k})| \ell(J_k) \leq 2\varepsilon/3 .$$

However, $g(s_k) = f(t_{j_k})$ since $s_k \in I_{j_k}$ so we have

$$\sum_{k=1}^{n} |f(s_k) - g(s_k)| \ell(J_k) = S(|f-g|, \mathcal{E}) \leq 2\varepsilon/3 .$$

Since $\mathcal{E}$ is also γ_2-fine, we have

$$\int_I |f-g| < S(|f-g|, \mathcal{E}) + \varepsilon/3 < \varepsilon .$$

9.2 Convergence Theorems

In this section we establish versions of the major convergence theorems (MCT, DCT) for the McShane integral. Due to the fact that the McShane integral is an absolute integral (Theorem 5), the conclusions of the convergence theorems for the McShane integral take on a stronger form than those for the gauge integral.

Throughout this section let I be a closed subinterval of $\mathbb{R}^*$, and let $f_k, f : I \to \mathbb{R}$. Some of the proofs or parts of the proofs for the convergence theorems are identical to the corresponding results for the gauge integral and will not be repeated, but references to the corresponding proofs will be given.

We begin with an analogue of Theorem 4.10.

Theorem 13. *Let $f : [a, \infty) = I \to \mathbb{R}$ be McShane integrable over I and set $F(t) = \int_a^t f, a \leq t < \infty$. Then $Var(F : I) = \int_I |f|$.*

Proof: Let $a = x_0 < x_1 < \ldots < x_n = b$ be a partition. Then

$$\sum_{j=1}^{n} |F(x_j) - F(x_{j-1})| \leq \sum_{j=1}^{n} \int_{x_{j-1}}^{x_j} |f| \leq \int_I |f|$$

so $Var(F : I) \leq \int_I |f|$.

For the reverse inequality, assume first that f is a step function $f = \sum_{j=1}^{n} t_j C_{A_j}$, where $A_j = [a_j, b_j]$ and the $\{A_j\}$ are non-overlapping, closed subintervals with union $J \subset I$. Then

$$\sum_{j=1}^{n} |F(b_j) - F(a_j)| = \sum_{j=1}^{n} \left| \int_{a_j}^{b_j} f \right|$$

$$= \sum_{j=1}^{n} |t_j| \ell(A_j) = \int_J |f| \leq \int_I |f|.$$

Therefore, $Var(F : J) \geq \int_I |f|$ and $Var(F : I) = \int_I |f|$ when f is a step function.

If f is McShane integrable over I, then for every k there is a step function g_k such that $\int_I |f - g_k| < 1/k$ (Theorem 12). Put $G_k(t) = \int_a^t g_k$. Then, by the first part of the proof,

$$|Var(G_k : I) - Var(F : I)| \leq Var(G_k - F : I) \leq \int_I |g_k - f| < 1/k$$

so $Var(G_k : I) \to Var(F : I)$. Since $\int_I |g_k| \to \int_I |f|$ and $Var(G_k : I) = \int_I |g_k|$, we have $Var(F : I) = \int_I |f|$.

This theorem and its proof should be contrasted to Theorem 4.10.

The result for arbitrary subintervals of $\mathbb{R}^*$ can be similarly formulated and established.

Analogous to Definition 5.1, we have

Definition 14. *Let $f_k : I \to \mathbb{R}$ be McShane integrable over I for each $k \in \mathbb{N}$. The sequence $\{f_k\}$ is uniformly McShane integrable over I if for every $\varepsilon > 0$ there is a gauge γ on I such that $|S(f_k, \mathcal{D}) - \int_I f_k| < \varepsilon$ for all k and all γ-fine free tagged partitions $\mathcal{D}$ of I.*

We next establish the analogue of Theorem 5.2 (7.8) for the McShane integral.

Theorem 15. *Let $f_k : I \to \mathbb{R}$. Suppose $\{f_k\}$ is uniformly McShane integrable over I and $\{f_k\}$ converges pointwise to the function f. Then f is McShane integrable over I and $\int_I |f_k - f| \to 0$.*

Proof: As in the proof of Theorem 5.2, $\lim \int_I f_k = L$ exists and f is McShane integrable with $\int_I f = L$. We do the remainder of the proof for the case when $I = [a, \infty]$; the other cases use the analogue of Theorem 13 for other subintervals.

Put $g_k = f_k - f$ so $g_k \to 0$ pointwise, and since f is McShane integrable, $\{g_k\}$ is uniformly McShane integrable. Let $\varepsilon > 0$. There exists a gauge γ

on I with $\gamma(t)$ bounded for every $t \in \mathbb{R}$ such that $|S(g_k, \mathcal{D}) - \int_I g_k| < \varepsilon$ for all k whenever $\mathcal{D}$ is a γ-fine free tagged partition of I. Fix such a partition $\mathcal{D} = \{(t_i, I_i) : 1 \le i \le m\}$ and assume that $t_1 = \infty$, $I_1 = [b, \infty]$. Let $b < \alpha$ and let $\{J_j : 1 \le j \le n\}$ be a partition of $[a, \alpha]$. Then $\mathcal{E} = \{(t_i, I_i \cap J_j) : 1 \le i \le m,\ 1 \le j \le n\}$ is γ-fine so Henstock's Lemma (3(vi)) implies

$$\left| \sum_{i=1}^{m} \sum_{j=1}^{n} \left\{ \int_{I_i \cap J_j} g_k - g_k(t_i)\ell(I_i \cap J_j) \right\} \right| \le \varepsilon .$$

Therefore,

(#)

$$\left| \sum_{j=1}^{n} \int_{J_j} g_k \right| \le \varepsilon + \left| \sum_{i=1}^{m} \sum_{j=1}^{n} g_k(t_i)\ell(I_i \cap J_j) \right|$$
$$= \varepsilon + \left| \sum_{i=2}^{m} g_k(t_i)\ell(I_i \cap [a, \alpha]) \right| \le \varepsilon + \max_{2 \le i \le m} |g_k(t_i)| \ell([a, b]) .$$

The last term on the right hand side of (#) can be made $< \varepsilon$ for large k since $g_k \to 0$ pointwise. Hence, (#) implies that $Var(G_k : [a, \alpha]) \le 2\varepsilon$ for large k. Since $\alpha > b$ is arbitrary, $Var(G_k : I) \le 2\varepsilon$ for large k. By Theorem 13, $\int_I |g_k| \le 2\varepsilon$ for large k as desired.

Theorem 15 should be compared to Theorem 5.2 (7.8) for the gauge integral. In particular, the much stronger conclusion should be noted; indeed, $\int_I |f_k - f| \to 0$ implies $\int_J (f_k - f) \to 0$ uniformly over all subintervals J of I (Theorem 7.8).

We now give an analogue of Theorem 5.4 for the McShane integral. Again, due to the fact that the McShane integral is an absolute integral and Theorem 15 allows a stronger conclusion than its counterpart for the gauge integral [Theorem 5.2], the statement and conclusion are somewhat different.

Theorem 16. *Let $f_k : I \to \mathbb{R}$ be McShane integrable over I for every k and suppose that $f = \sum_{k=1}^{\infty} f_k$ pointwise on I with $\sum_{k=1}^{\infty} \int_I |f_k| < \infty$. If $s_n = \sum_{k=1}^{n} f_k$, then*

(i) *$\{s_n\}$ is uniformly McShane integrable over I,*
(ii) *f is McShane integrable over I with $\int_I f = \sum_{k=1}^{\infty} \int_I f_k$ and*
(iii) *$\int_I |s_n - f| = \int_I |\sum_{k=n+1}^{\infty} f_k| \to 0$.*

The proof of this result is essentially the same as the proof of Theorem 5.4 so we do not repeat the argument. However, a few remarks are in order. Since the McShane integral is an absolute integral, we do not need to assume that the $\{f_k\}$ are non-negative as in Theorem 5.4, and we then replace the assumption that $\sum_{k=1}^{\infty} \int_I f_k < \infty$ with $\sum_{k=1}^{\infty} \int_I |f_k| < \infty$. The proof of Theorem 16 (i) then proceeds as the proof of Theorem 5.4 (i) except that in the estimation of the term T_2 the absolute integrability of f_k is used. Conclusions (ii) and (iii) of Theorem 16 now follow from part (i) and Theorem 15. Since

$$\left| \sum_{k=1}^{n} \int_I f_k - \int_I f \right| = \left| \sum_{k=n+1}^{\infty} \int_I f_k \right| \le \sum_{k=n+1}^{\infty} \int_I |f_k| ,$$

the conclusion in Theorem 16 (iii) is stronger than the conclusion in Theorem 5.4 (iii).

As in Theorem 5.5 the MCT now follows readily from Theorem 16.

Theorem 17 (*MCT*). *Let $f_k : I \to \mathbb{R}$ be McShane integrable over I and suppose that $f_k(t) \uparrow f(t) \in \mathbb{R}$ for every $t \in I$. If $\sup_k \int_I f_k < \infty$, then*

(i) *$\{f_k\}$ is uniformly McShane integrable over I,*
(ii) *f is McShane integrable over I and*
(iii) $\lim \int_I f_k = \int_I f (= \int_I (\lim f_k))$.

Proof: As before set $f_0 = 0$ and $g_k = f_k - f_{k-1}$ for $k \ge 1$. Then $g_k \ge 0$, $\sum_{k=1}^{n} g_k = f_n \to f$ pointwise and $\sum_{k=1}^{\infty} \int_I g_k = \lim_n \sum_{k=1}^{n} \int_I (f_k - f_{k-1}) = \lim_n \int_I f_n = \sup_n \int_I f_n < \infty$. Hence, Theorem 16 is applicable and gives the result.

The DCT for the McShane integral can be derived exactly as in Chapter 5 for the gauge integral. We can, however, obtain a stronger conclusion in the DCT for the McShane integral. We repeat the essential steps.

Definition 18. *Let $f_k : I \to \mathbb{R}$. The sequence $\{f_k\}$ is uniformly McShane Cauchy over I if for every $\varepsilon > 0$ there exist a gauge γ on I and an N such that if $i, j \ge N$, then $|S(f_i, \mathcal{D}) - S(f_j, \mathcal{D})| < \varepsilon$ whenever $\mathcal{D}$ is a γ-fine free tagged partition of I.*

Then exactly as in Proposition 5.7, we have

Proposition 19. *Let $f_k : I \to \mathbb{R}$ be McShane integrable over I. Then $\{f_k\}$ is uniformly McShane Cauchy over I if and only if $\{f_k\}$ is uniformly McShane integrable over I and $\lim \int_I f_k$ exists.*

Theorem 20 (*DCT*). *Let $f_k, g : I \to \mathbb{R}$ be McShane integrable over I with $|f_k| \le g$ on I. If $f_k \to f$ pointwise on I, then*

(i) *$\{f_k\}$ is uniformly McShane integrable over I,*
(ii) *f is McShane integrable over I and*
(iii) $\int_I |f_k - f| \to 0$.

The proof of Theorem 20 (i) proceeds as the proof of Theorem 5.8 (i). Parts (ii) and (iii) then follow immediately from Theorem 15.

A few remarks are in order. First, the domination assumption $|f_k| \le g$ implies $|f_k - f_j| \le 2g$ as in Theorem 5.8; however, the domination assumption $|f_k - f_j| \le g$ in Theorem 5.8 allows for conditionally convergent integrals in the case of the gauge integral, whereas the assumption $|f_k| \le g$ does not. Next, since $|\int_I f_k - \int_I f| \le \int_I |f_k - f|$, the conclusion (iii) in Theorem 20 implies conclusion (iii) in Theorem 5.8.

The continuity and differentiation results for integrals containing parameters given in Theorems 5.10 and 5.12 can be obtained in exactly the same way for the McShane integral. We do not repeat the statements.

We next establish a result for "improper" McShane integrals.

Theorem 21. *Let $f : [a, b] \to \mathbb{R}$ and let $a \le b_k < b, b_k \uparrow b$. Suppose that f is McShane integrable over $[a, b_k]$ for every k. Then f is McShane integrable over $[a, b]$ if and only if $\sup\{\int_a^{b_k} |f| : k\} < \infty$. In this case, $\lim \int_a^{b_k} f = L$ exists and $\int_a^b f = L$.*

Proof: If f is integrable over $[a, b]$, $\int_a^b |f| \ge \int_a^{b_k} |f|$ for every k.

Conversely, assume $\sup\{\int_a^{b_k} |f| : k\} < \infty$. First, consider the case when $f \ge 0$ and set $f_k = C_{[a,b_k]} f$. Then f_k is integrable over $[a, b]$, $\int_a^{b_k} f = \int_a^b f_k$ and $f_k \uparrow f$. The MCT implies that f is integrable over $[a, b]$ with $\int_a^{b_k} f \uparrow \int_a^b f$. If f is integrable over $[a, b]$, both f^+ and f^- are integrable over $[a, b]$ so the result just established can be applied to both f^+ and f^-.

Theorem 21 should be compared to Theorem 3.4 for the gauge integral. A comparison theorem for the McShane integral is given in Exercise 15.

Finally, we conclude this section by establishing a result for the indefinite McShane integral.

Definition 22. *Let $g : [a, b] \to \mathbb{R}$. Then g is absolutely continuous if for every $\varepsilon > 0$ there exists a $\delta > 0$ such that if $\{[a_k, b_k] : 1 \le k \le n\}$ is any finite family of non-overlapping subintervals of $[a, b]$ with $\sum_{k=1}^n (b_k - a_k) < \delta$, then $\sum_{k=1}^n |g(b_k) - g(a_k)| < \varepsilon$.*

For example, any function satisfying a Lipschitz condition on $[a,b]$ is obviously absolutely continuous. [See Exercise 18 for the converse.]

Proposition 23. *Let $f : [a,b] \to \mathbb{R}$ be absolutely continuous. Then f is uniformly continuous and of bounded variation on $[a,b]$.*

Proof: f is obviously uniformly continuous. Let $\varepsilon = 1$ and δ be as in Definition 22. Partition $[a,b]$ by $\mathcal{P} = \{a = x_0 < x_1 < \cdots < x_n = b\}$ with $x_k - x_{k-1} < \delta, 1 \le k \le n$. Then $Var(f : [x_{k-1}, x_k]) \le 1$ so $Var(f : [a,b]) \le n$.

The converse of Proposition 23 is false; the standard counter-example is the so-called Cantor function (see [Sw1] 4.4.9).

Theorem 24. *Let $f : [a,b] \to \mathbb{R}$ be McShane integrable over $[a,b]$ and set $F(t) = \int_a^t f, a \le t \le b$. Then F is absolutely continuous on $[a,b]$.*

Proof: Let $\varepsilon > 0$. Put $f_k = |f| \wedge k$. Then f_k is integrable (Corollary 8) and $f_k \uparrow |f|$. By MCT $\int_a^b f_k \uparrow \int_a^b |f|$. Pick k such that $0 \le \int_a^b (|f| - f_k) < \varepsilon/2$ and set $\delta = \varepsilon/2k$. Assume that $\{[a_i, b_i] : 1 \le i \le m\}$ is a sequence of non-overlapping subintervals of $[a,b]$ the sum of whose lengths is less than δ. Then

$$\sum_{i=1}^m |F(b_i) - F(a_i)| \le \sum_{i=1}^m \int_{a_i}^{b_i} |f| = \sum_{i=1}^m \int_{a_i}^{b_i} (|f| - f_k) + \sum_{i=1}^m \int_{a_i}^{b_i} f_k$$

$$\le \int_a^b (|f| - f_k) + k \sum_{i=1}^m (b_i - a_i) < \varepsilon\,.$$

Theorem 24 should be compared to Theorem 4.9 for the gauge integral.

9.3 Integrability of Products and Integration by Parts

In this section we consider the integrability of the product of two McShane integrable functions (see Exercise 24). As an application we derive an integration by parts formula for the McShane integral. The results of this section are not used later and may be skipped.

Throughout this section let I be a closed subinterval of $\mathbb{R}^*$ and let $f, g : I \to \mathbb{R}$. We begin with a preliminary result on products.

Proposition 25. *Let f be bounded and McShane integrable over I. Then f^n is McShane integrable over I for every $n \in \mathbb{N}$.*

Proof: The function $h(t) = t^n$ satisfies a Lipschitz condition on every bounded interval (Exercise 6). Hence, $h \circ f = f^n$ is McShane integrable over I by Theorem 7.

Corollary 26. *Let f, g be bounded and McShane integrable over I. Then fg is McShane integrable over I.*

Proof: Since $2fg = (f+g)^2 - f^2 - g^2$, the result follows from Proposition 25.

To remove the boundedness hypothesis in Corollary 26, we establish 2 lemmas.

Lemma 27. *Let f be non-negative and McShane integrable over I. Then there exists an increasing sequence of non-negative, bounded, McShane integrable functions $\{f_k\}$ such that $f_k \uparrow f$ on I.*

Proof: Set $f_k = f \wedge k$ and apply Corollary 8.

Lemma 28. *Let f, g be non-negative and McShane integrable over I. If there exists a McShane integrable function $h : I \to \mathbb{R}$ such that $fg \le h$ on I, then fg is McShane integrable over I.*

Proof: Let $\{f_k\}, \{g_k\}$ be non-negative, bounded, McShane integrable and such that $f_k \uparrow f, g_k \uparrow g$ on I (Lemma 27). Then $f_k g_k$ is McShane integrable (Corollary 26), $f_k g_k \uparrow fg$ and $\int_I f_k g_k \le \int_I h$. The MCT gives the result.

Finally, we have our basic result.

Theorem 29. *Let f, g be McShane integrable over I with g bounded on I. Then fg is McShane integrable over I.*

Proof: Since $fg = (f^+ - f^-)(g^+ - g^-)$, we may assume that f and g are non-negative. If $|g(t)| \le M$ for all $t \in I$, $fg \le Mf$ and the result follows from Lemma 28.

Theorem 29 should be compared with Example 4.14 for the gauge integral. See Exercise 25 for an improvement in Theorem 29 and see Exercise 24 for the boundedness assumption.

We can now use Theorem 29 to derive an integration by parts formula for the McShane integral.

Theorem 30 (*Integration by Parts*). *Let $f, g : [a, b] \to \mathbb{R}$ be differentiable in $[a, b]$ and assume that f' and g' are McShane integrable over $[a, b]$. Then $f'g$ and fg' are McShane integrable over $[a, b]$ with $\int_a^b fg' = fg|_a^b - \int_a^b f'g$.*

Proof: f and g are bounded and continuous and are, therefore, McShane integrable. By Theorem 29 fg' and $f'g$ are McShane integrable. Since $(fg)' = f'g + fg', (fg)'$ is McShane integrable and Theorem 10 gives $\int_a^b (fg)' = fg|_a^b = \int_a^b f'g + \int_a^b fg'$.

Theorem 30 should be compared with Corollary 2.3.

Similarly, we can use Theorem 30 to derive a result on integration by substitution or change of variable.

Theorem 31 (*Integration by Substitution*). *Let $f : [c, d] \to \mathbb{R}$ be continuous and let $g : [a, b] \to [c, d]$ be differentiable on $[a, b]$ with g' McShane integrable over $[a, b]$. Then $(f \circ g)g'$ is McShane integrable over $[a, b]$ with $\int_{g(a)}^{g(b)} f = \int_a^b (f \circ g)g'$.*

Proof: Set $F(u) = \int_{g(a)}^{u} f$ for $c \le u \le d$. Then $F' = f$ on $[c, d]$ by Theorem 11. By Theorem 29 $(f \circ g)g'$ is integrable since $f \circ g$ is continuous on $[a, b]$. But, $(F \circ g)' = (f \circ g)g'$ so Theorem 10 gives $\int_a^b (f \circ g)g' = F \circ g|_a^b = \int_{g(a)}^{g(b)} f$.

Theorem 31 should be compared to Theorem 1.6 for the gauge integral.

9.4 More General Convergence Theorems

In this section we give a more general form of the MCT which is very useful in many applications. We also use the generalization to establish a version of Fatou's Lemma and a more general form of the DCT. Recall that a subset $E \subset \mathbb{R}$ is *null* or is a null set if for every $\varepsilon > 0$ there is a cover of E by open intervals $\{G_j\}$ such that $\sum_{j=1}^{\infty} \ell(G_j) < \varepsilon$ (Definition 1.9).

Theorem 32. *E is null if and only if C_E is McShane integrable over $\mathbb{R}$ with $\int_{\mathbb{R}} C_E = 0$.*

Proof: Suppose E is null and let $\varepsilon > 0$. Let $\{G_j\}$ be a sequence of open intervals covering E with $\sum_{j=1}^{\infty} \ell(G_j) < \varepsilon$. Put $s_n = C_{G_1} \vee \cdots \vee C_{G_n}$. Since $s_n \uparrow, h = \lim s_n$ exists. Note that $0 \le h \le 1$ and $C_E \le h$. Also, $s_n \le \sum_{k=1}^{n} C_{G_k}$ implies that $\int_{\mathbb{R}} s_n \le \sum_{k=1}^{n} \ell(G_k) \le \sum_{k=1}^{\infty} \ell(G_k) < \varepsilon$. Hence, the MCT implies that h is integrable with $\int_{\mathbb{R}} h < \varepsilon$. Since $0 \le C_E \le h, C_E$ is integrable and $\int_{\mathbb{R}} C_E = 0$ by Theorem 3 (iv).

For the converse let I be an arbitrary closed, bounded interval and put $A = I \cap E$. It suffices to show that A is null. Note that C_A is integrable with $\int_{\mathbb{R}} C_A = \int_I C_A = 0$. There exists a gauge γ on $\mathbb{R}$ such that $|S(C_A, \mathcal{D})| < \varepsilon$ whenever $\mathcal{D}$ is a γ-fine free tagged partition of $\mathbb{R}$. Let $\{(t_k, I_k) : k \in M\}$ be the cover from Lemma 6.3 relative to A, I, ε, γ. For $n \in \mathbb{N}$, let $M_n = \{k \in M : 1 \le k \le n\}$. Since $\{(t_k, I_k) : k \in M_n\} << \gamma$, by Henstock's Lemma we have

$$\left| \sum_{k \in M_n} C_A(t_k)\ell(I_k) \right| = \sum_{k \in M_n} \ell(I_k) < \varepsilon$$

for every n. Hence, $\sum_{k\in M} \ell(I_k) \leq \varepsilon$. Each I_k can be expanded to an open interval G_k containing I_k with $\sum_{k\in M} \ell(G_k) < 2\varepsilon$. Since $A \subset \bigcup_{k\in M} I_k \subset \bigcup_{k\in M} G_k$, this means that A is null.

Theorem 32 and particularly its proof should be compared with Example 1.10 and Corollary 6.6 for the gauge integral.

Recall that property P concerning the points of a subset $A \subset \mathbb{R}$ is said to hold almost everywhere (a.e.) in A if the property P holds for all points in A except those in a null set.

For the more general version of MCT we first require some preliminary results.

Lemma 33. *Let $f_k : I \to \mathbb{R}$ be non-negative and McShane integrable over I with $f_k \leq f_{k+1}$ on I. Let $f(t)$ be the limit of $\{f_k(t)\}$ in the extended reals, $\mathbb{R}^*$. If $\sup_k \int_I f_k = M < \infty$, then f is finite a.e. in I.*

Proof: Let $E = \{t \in I : f(t) = \infty\}$. For each j and k, f_k/j is integrable so $(f_k/j) \wedge 1$ is integrable (Corollary 8) with $\int_I (f_k/j) \wedge 1 \leq \int_I (f_k/j) \leq M/j$.

For each $j (f_k/j) \wedge 1 \uparrow (f/j) \wedge 1$ as $k \to \infty$ so the MCT implies

$$(*)\ \lim_k \int_I (f_k/j) \wedge 1 = \int_I (f/j) \wedge 1 \leq M/j.$$

Now $f \geq 0$ so $(f/j) \wedge 1 \downarrow$ and $\lim_j \int_I (f/j) \wedge 1 = 0$ by (*). If $t \in I \backslash E$, $(f(t)/j) \wedge 1 = f(t)/j$ for large j so $(f(t)/j) \wedge 1 \to 0$; if $t \in E$, $(f(t)/j) \wedge 1 = 1$ so $(f(t)/j) \wedge 1 \to 1$. That is, $(f/j) \wedge 1 \to C_E$ pointwise so by the MCT C_E is integrable with $\int_I C_E = 0$. Theorem 32 now gives the result.

Lemma 34. *Let $f : I \to \mathbb{R}$ be McShane integrable over I and $g : I \to \mathbb{R}$ be such that $f = g$ a.e. in I. Then g is McShane integrable over I with $\int_I f = \int_I g$.*

Proof: Put $h = f - g$ so $h = 0$ a.e. in I. It suffices to show that h is integrable with $\int_I h = 0$. Let $E = \{t : h(t) \neq 0\}$.

Put $h_n = |h| \wedge n$. Then $h_n \leq nC_E$ implies h_n is integrable with $\int_I h_n = 0$ (Exercise 4). Since $h_n \uparrow |h|$, the MCT implies that $|h|$ is integrable with $\int_I |h| = 0$. Thus, h is integrable with $\int_I h = 0$ (Exercise 4).

This lemma and its proof should be compared with Example 1.10 (Exercise 2.22) for the gauge integral.

We now have a more general form for Theorem 16.

Theorem 35. *Let each $f_k : I \to \mathbb{R}$ be McShane integrable over I with $\sum_{k=1}^{\infty} \int_I |f_k| < \infty$. There exists a McShane integrable function $f : I \to \mathbb{R}$ such that if $s_n = \sum_{k=1}^{n} f_k$, then $s_n \to f$ a.e. and $\int_I |s_n - f| \to 0$.*

Proof: Set $g_n = \sum_{k=1}^{n} |f_k|$. Then $g_n \uparrow \sum_{k=1}^{\infty} |f_k|$ and since

$$\int_I g_n = \sum_{k=1}^{n} \int_I |f_k| \le \sum_{k=1}^{\infty} \int_I |f_k| < \infty ,$$

Lemma 33 implies that $\sum_{k=1}^{\infty} |f_k|$ converges in $\mathbb{R}$ a.e. in I. Let $E = \{t : \sum_{k=1}^{\infty} |f_k(t)| < \infty\}$ and set $h_k = C_E f_k$, $f(t) = \sum_{k=1}^{\infty} f_k(t)$ for $t \in E$ and $f(t) = 0$ for $t \notin E$. Then $f_k = h_k$ a.e., $\int_I f_k = \int_I h_k$ (Lemma 34), $\sum_{k=1}^{n} h_k \to f$ pointwise on I and $\sum_{k=1}^{\infty} \int_I |h_k| < \infty$. Theorem 16 now applies so f is McShane integrable with

$$\int_I \left| \sum_{k=1}^{n} h_k - f \right| = \int_I \left| \sum_{k=1}^{n} f_k - f \right| \to 0 .$$

Our general form of the MCT now follows easily from Theorem 35.

Theorem 36 (*MCT*). *Suppose each $f_k : I \to \mathbb{R}$ is McShane integrable over I with $f_k \le f_{k+1}$ and $\sup_k \int_I f_k < \infty$. Then there exists a McShane integrable function $f : I \to \mathbb{R}$ such that $f_k \to f$ a.e. and $\int_I f_k \uparrow \int_I f$.*

Proof: Set $f_0 = 0$ and $g_k = f_k - f_{k-1}$ for $k \ge 1$. Then $g_k \ge 0$ and

$$\sum_{k=1}^{\infty} \int_I |g_k| = \lim_n \sum_{k=1}^{n} \int_I (f_k - f_{k-1}) = \lim_n \int_I f_n < \infty .$$

Theorem 35 implies there is a McShane integrable function $f : I \to \mathbb{R}$ such that $\sum_{k=1}^{\infty} g_k = \lim_n f_n = f$ a.e. and $\sum_{k=1}^{\infty} \int_I g_k = \lim_n \int_I f_n = \int_I f$.

We can now employ Theorem 36 to obtain a general form of Fatou's Lemma.

Lemma 37 (*Fatou*). *Let $f_k : I \to [0, \infty)$ be McShane integrable over I. Suppose that $\underline{\lim} \int_I f_k < \infty$. Then $\underline{\lim} f_k = f$ is finite a.e. and if $g : I \to \mathbb{R}$ is such that $f = g$ a.e. in I, then g is McShane integrable over I with $\int_I g \le \underline{\lim} \int_I f_k$.*

Proof: Put $g_k = f_1 \wedge \cdots \wedge f_k$. Then $g_k \ge 0, g_k \uparrow \underline{\lim} f_k$, $0 \le g_k \le f_k$ and each g_k is integrable (Corollary 6). Therefore, $\lim \int_I g_k \le \underline{\lim} \int_I f_k < \infty$. By Lemma 33 $\underline{\lim} f_k$ is finite a.e. and by Theorem 36 $\int_I g = \lim \int_I g_k \le \underline{\lim} \int_I f_k$.

Remark 38. *The non-negativity assumption is important* (*Exercise* 20) *and strict inequality is possible* (*Exercise* 21). *If extended real-valued functions are allowed in the integral, the conclusion in Fatou's Lemma is usually written as $\int_I \underline{\lim} f_k \le \underline{\lim} \int_I f_k$.*

Finally, we obtain a general form of the DCT involving convergence a.e.

Theorem 39 (*DCT*). *Let $f_k, f, g : I \to \mathbb{R}$ with f_k, g McShane integrable over I. Assume $|f_k| \le g$ a.e. for each k and $f_k \to f$ a.e. in I. Then f is McShane integrable over I with $\int_I |f_k - f| \to 0$.*

Proof: $g - f_k \ge 0$ a.e. and $\underline{\lim} \int_I (g - f_k) = \int_I g - \overline{\lim} \int_I f_k < \infty$ since $|\int_I f_k| \le \int_I |f_k| \le \int_I g$. Fatou's Lemma implies that f is integrable with

$$\int_I \underline{\lim}(g - f_k) = \int_I (g - f) = \int_I g - \int_I f \le \underline{\lim} \int_I (g - f_k) = \int_I g - \overline{\lim} \int_I f_k$$

or $\overline{\lim} \int_I f_k \ge \int_I f$.

Similarly, $f_k + g \ge 0$ a.e. and $\underline{\lim} \int_I (f_k + g) = \underline{\lim} \int_I f_k + \int_I g < \infty$ implies by Fatou's Lemma $\int_I (f + g) \le \underline{\lim} \int_I f_k + \int_I g$ or $\underline{\lim} \int_I f_k \ge \int_I f$. Hence $\lim \int_I f_k$ exists and equals $\int_I f$.

Note that the conclusion of Theorem 39 is not as sharp as the conclusion of the version of the DCT given by Theorem 20 (i).

9.5 The Space of McShane Integrable Functions

In this section we consider the space of McShane integrable functions. We show that there is a natural semi-norm defined on the space and in contrast to the Alexiewicz semi-norm defined on the space of gauge integrable functions in Chapter 7, the space of McShane integrable functions is complete under this semi-norm. This section requires only a very basic knowledge of (semi-) normed spaces and the general MCT and DCT given in Sec. 9.4.

Let I be a closed interval in $\mathbb{R}^*$ and let $\mathcal{M}(I)$ be the space of all functions which are McShane integrable over I. Then $\mathcal{M}(I)$ is a (real) vector space and has a very natural semi-norm defined on it via $\|f\|_1 = \int_I |f|$; $\| \|_1$ is often called the L^1-*norm*. Note that $\| \|_1 = 0$ if and only if $f = 0$ a.e. in I (Lemma 34 and Exercise 19); it is often the case that functions in $\mathcal{M}(I)$ are identified if they are equal a.e. in I so $\| \|_1$ can then be considered to be a norm on $\mathcal{M}(I)$.

In contrast to the situation with the space of gauge integrable functions under the Alexiewicz norm (7.1), we show that $(\mathcal{M}(I), \| \|_1)$ is complete.

Theorem 40. $(\mathcal{M}(I), \| \|_1)$ *is complete.*

Proof: Let $\{f_k\} \subset \mathcal{M}(I)$ be Cauchy with respect to $\| \|_1$. It suffices to show that there exist a subsequence $\{f_{n_k}\}$ of $\{f_k\}$ and a function $f \in \mathcal{M}(I)$ such that $\|f_{n_k} - f\|_1 \to 0$. Pick a subsequence n_k such that $\|f_{n_{k+1}} - f_{n_k}\|_1 < 1/2^k$

and put $g_k = f_{n_{k+1}} - f_{n_k}$. Then $\sum_{k=1}^{\infty} \int_I |g_k| < \infty$ so by Theorem 35 there exists $g \in \mathcal{M}(I)$ such that $\sum_{k=1}^{\infty} g_k = g$ a.e. and

$$\int_I \left| \sum_{i=1}^{j} g_i - g \right| = \int_I |f_{n_{j+1}} - f_{n_1} - g| \to 0 .$$

Thus, if $f = f_{n_1} + g$, $f \in \mathcal{M}(I)$ and $\|f_{n_j} - f\|_1 \to 0$.

Note from Theorem 12 that the subspace of step functions is dense in $\mathcal{M}(I)$ with respect to $\| \|_1$.

9.6 Multiple Integrals and Fubini's Theorem

As was the case for the gauge integral, except for the usual notational difficulties, there is no problem in defining the McShane integral for functions defined on intervals in $\mathbb{R}^n$. In this section we give the definition. Most of the basic results are analogous to those for the McShane integral in $\mathbb{R}^1$ and the versions of the Fubini Theorems for the gauge integral carry forward to the McShane integral.

Let $\mathbb{R}^{*n} = \mathbb{R}^* \times \cdots \times \mathbb{R}^*$ where there are n factors in the product. An *interval* I is a product $I_1 \times \cdots \times I_n$ where each I_k is an interval in $\mathbb{R}^*$; we say that I is a closed (open) interval if each I_k is closed (open). The *interior* of I, written I°, is the product of the interiors of the I_k. The *volume* of an interval I, denoted $v(I)$, is the product $v(I) = \prod_{k=1}^{n} \ell(I_k)$, where we continue to use the convention that $0 \cdot \infty = 0$. [In $\mathbb{R}^2$, the term *area* would be more appropriate.]

If I is a closed interval, a *partition* of I is a finite collection of closed subintervals of I, $\{I_k : 1 \le k \le m\}$, such that $I_k^\circ \cap I_j^\circ = \emptyset$ if $k \ne j$ [i.e., the $\{I_k\}$ do not overlap and $I = \bigcup_{k=1}^{m} I_k$]. A *free tagged partition* $\mathcal{D}$ of I is a finite collection of pairs $\{(x_i, I_i) : 1 \le i \le m\}$ such that $\{I_i : 1 \le i \le m\}$ is a partition of I and $x_i \in I$ for $1 \le i \le m$; we call the I_i the subintervals of $\mathcal{D}$ and call x_i the *tag* associated with I_i. Note again that in a free tagged partition the tag x_i is not required to belong to its associated subinterval I_i. A *gauge* γ on I is a function defined on I such that $\gamma(x)$ is an open interval containing x for every $x \in I$. A free tagged partition $\mathcal{D} = \{(x_i, I_i) : 1 \le i \le m\}$ is γ-*fine* if $I_i \subset \gamma(x_i)$ for $1 \le i \le m$; we write $\mathcal{D} << \gamma$ as before.

If f is a function defined on an interval I, we assume that f vanishes at all infinite points. If $f : I \to \mathbb{R}$ and $\mathcal{D} = \{(x_i, I_i) : 1 \le i \le m\}$ is a free tagged partition of I, the *Riemann sum* of f with respect to $\mathcal{D}$ is defined to be $S(f, \mathcal{D}) = \sum_{i=1}^{m} f(x_i) v(I_i)$ [again we assume $0 \cdot \infty = 0$].

Definition 41. *Let I be a closed subinterval of $\mathbb{R}^{*n}$ and $f : I \to \mathbb{R}$. Then f is McShane integrable over I if there exists $A \in \mathbb{R}$ such that for every $\varepsilon > 0$ there exists a gauge γ on I such that $|S(f, \mathcal{D}) - A| < \varepsilon$ whenever $\mathcal{D}$ is a γ-fine free tagged partition of I.*

From Corollary 8.2 it follows that Definition 41 makes sense and as before the number A is unique, is called the McShane integral of f over I and is denoted by $A = \int_I f$, etc.

The basic properties of the integral such as linearity, additivity, positivity, etc., carry forward to the McShane integral over $\mathbb{R}^n$. We do not repeat the statements or proofs (see Theorems 2 and 3). In particular, the analogue of Theorem 5 on absolute integrability holds in $\mathbb{R}^{*n}$.

The proof in Example 8.4 is valid for free tagged partitions so it follows that any step function in $\mathbb{R}^n$ is McShane integrable. Also, Theorem 8.7 is valid for the McShane integral and gives a large class of functions which are McShane integrable.

The junior grade versions of Fubini's Theorem (Tonelli's Theorem) given in Chapter 8 (Theorems 8 and 9) for the gauge integral hold with no change for the McShane integral. The proofs given in Chapter 8 utilize the integrability of step functions, the Bounded Convergence Theorem and MCT, all of which are valid for the McShane integral. Again, we do not repeat the statements or proofs but refer the reader to Chapter 8.

Similarly, the general form of Fubini's Theorem given in Chapter 8 (Theorem 8.13) for the gauge integral holds for the McShane integral. This version requires the analogue of Lemma 8.1 for free tagged partitions and the use of null sets and the analogue of Corollary 6.6 given in Theorem 32. It should be pointed out that Theorem 8.13 for the gauge integral admits conditionally integrable functions whereas its analogue for the McShane integral does not.

Notes/Remarks

As was the case for the gauge integral, there is a stronger form of Part 2 of the FTC than that given in Theorem 11. If $f : [a, b] \to \mathbb{R}$ is McShane integrable over $[a, b]$ and $F(t) = \int_a^t f, a \le t \le b$, is the indefinite integral of f, then F is differentiable a.e. in $[a, b]$ with $F' = f$ a.e. This follows from the result in Appendix 3 for the gauge integral since any McShane integrable function is gauge integrable.

There is also a converse to Theorem 24 which gives a characterization of indefinite McShane integrals. If $F : [a, b] \to \mathbb{R}$ is absolutely continuous on $[a, b]$, then F is differentiable a.e. in $[a, b]$ with $F(t) - F(a) = \int_a^t F'$ [here F' is defined to be 0 where F is not differentiable]. This is a well-known result for the Lebesgue integral [see [Sw1] 4.4.7] and is sometimes referred to as the Radon-Nikodym Theorem for the Lebesgue integral. [See Appendix 4.]

Theorem 40 on the completeness of the space of McShane integrable functions is usually referred to as the Riesz-Fischer Theorem for the Lebesgue integral ([Sw1] 3.5.1).

There is in some sense a converse to the result in Theorem 29. If $g : I \to \mathbb{R}$ is measurable (Lebesgue) and fg is McShane integrable over I for every $f \in \mathcal{M}(I)$, then g is equal a.e. to a bounded measurable function (such functions are said to be *essentially bounded*). This solves what might be called the "multiplier problem" for the McShane integral (see the Notes/Remarks section of Chapter 4). This is a well-known result for the Lebesgue integral ([Sw1] Exer. 3.2.21), and since the McShane and Lebesgue integrals are equivalent (Appendix 4), this means the same result holds for the McShane integral.

Exercises

1. Let $\mathcal{D} = \{(t_i, I_i) : 1 \le i \le m\}$ be a free tagged partition of I and let $\mathcal{J} = \{J_j : 1 \le j \le n\}$ be a partition of I. Show $\mathcal{D}' = \{(t_i, I_i \cap J_j) : 1 \le i \le m, 1 \le j \le n, I_i^\circ \cap J_j^\circ \ne \emptyset\}$ is a free tagged partition of I and $S(f, \mathcal{D}) = S(f, \mathcal{D}')$. Moreover, if γ is a gauge on I and $\mathcal{D} << \gamma$, then $\mathcal{D}' << \gamma$. Does a similar result hold for (gauge) tagged partitions?
2. If $f : [a, b] \to \mathbb{R}$ is increasing, show that f is McShane integrable over $[a, b]$. [Hint: Divide $[a, b]$ into n equal length subintervals, $a = x_0 < x_1 < \cdots < x_n = b$. Put $I_k = [x_{k-1}, x_k]$, $M_k = f(x_k)$, $m_k = f(x_{k-1})$, $h = \sum M_k C_{I_k}$, $g = \sum m_k C_{I_k}$ and use Theorem 3 (iv).]
3. (Linear Change of Variable) Let $f : [a, b] \to \mathbb{R}$ be McShane integrable over $[a, b]$ and $h \in \mathbb{R}$. Define $f_h : [a + h, b + h] \to \mathbb{R}$ by $f_h(t) = f(t - h)$. Show that f_h is McShane integrable with $\int_a^b f = \int_{a+h}^{b+h} f_h$.
4. Let $f, h : I \to \mathbb{R}$. Assume that $|f| \le h$ on I and h is integrable over I with $\int_I h = 0$.
Show that f is integrable with $\int_I |f| = 0$.
5. Let $f : I \to \mathbb{R}$. Show that f is integrable over I if and only if f^+ and f^- are integrable over I.
6. Let $f : I \to \mathbb{R}$. Suppose f is differentiable on I and f' is bounded on I. Show that f satisfies a Lipschitz condition on I.
7. Show that $f(t) = \sqrt{t}, 0 \le t \le 1$, does not satisfy a Lipschitz condition.
8. Let $c, s, t \in \mathbb{R}$. Show $|c \wedge t - c \wedge s| \le |t - s|$ and $|c \vee t - c \vee s| \le |t - s|$.
9. Assume the notation in Theorem 7 and let L be a Lipschitz constant for g. Show that $\int g \circ f \le L \int |f|$.
10. Give a counter-example to Corollary 8 for the gauge integral.

11. Let $f : [a, \infty) \to \mathbb{R}$ be McShane integrable. Show that $\lim_{b\to\infty} \int_b^\infty |f| = 0$. [Hint: Pick γ such that $\gamma(t)$ is bounded for $t \in \mathbb{R}$ and $|S(|f|, \mathcal{D}) - \int_a^\infty |f|| < \varepsilon$ whenever $\mathcal{D} << \gamma$. Fix such a $\mathcal{D} = \{(t_i, I_i) : 1 \le i \le m\}$ with $t_1 = \infty, I_1 = [b, \infty]$. Consider $\int_c^\infty |f|$ for $c > b$.]
12. Can you formulate and prove a version of Exercise 11 for a sequence of uniformly McShane integrable functions?
13. Let $f : I \to \mathbb{R}$ be McShane integrable. Show that $\lim_{\ell(J)\to 0} \int_J |f| = 0$. [Hint: Pick γ such that $|S(|f|, \mathcal{D}) - \int_I |f|| < \varepsilon$ when $\mathcal{D} << \gamma$. Fix such a $\mathcal{D} = \{(t_i, I_i) : 1 \le i \le m\}$ and put $M = \max\{|f(t_i)| : 1 \le i \le m\}$. Let J be a subinterval of I. Consider $\mathcal{E} = \{(t_i, I_i \cap J) : 1 \le i \le m\}$ and use Henstock's Lemma to see how to choose δ so that $\ell(J) < \delta$ implies $\int_J |f| \le 2\varepsilon$.]
14. For what values of p is t^p McShane integrable over $[0, 1]$?
15. (Comparison Theorem) Let $f, g : [a, b] \to \mathbb{R}$, $a < b_k < b$, $b_k \uparrow b$. Suppose f is McShane integrable over $[a, b_k]$ for every k and g is McShane integrable over $[a, b]$ with $|f| \le g$ on $[a, b]$. Show that f is McShane integrable over $[a, b]$.
16. Establish analogues of Theorem 21 and Exercise 15 for unbounded intervals (improper integrals).
17. Use Exercise 16 to determine the values of p for which t^p is McShane integrable over $[1, \infty)$.
18. Show that $g(t) = t^{1/3}, 0 \le t \le 1$, is absolutely continuous but does not satisfy a Lipschitz condition.

19*. Let $f : I \to [0, \infty)$ and suppose f is integrable over I with $\int_I f = 0$. Show that $f = 0$ a.e. in I. [Hint: Let $E = \{t : f(t) > 0\}$ and put $f_k = (kf) \wedge C_E$ (Exercise 6.10).]

20*. Show that the non-negativity assumption in Fatou's Lemma is important.

21*. Show that strict inequality can hold in Fatou's Lemma. [Hint: Consider $f_k = C_{(0,2)}$ for k odd and $f_k = C_{(1,3)}$ for k even.]

22*. Let $f : \mathbb{R} \to \mathbb{R}$ be McShane integrable over $\mathbb{R}$. Let $\mathcal{M}$ be the class of Lebesgue measurable subsets of $\mathbb{R}$ and define $\nu : \mathcal{M} \to \mathbb{R}$ by $\nu(E) = \int_E f$ (Chapter 6). Show that ν is countably additive (compare with 6.9 and Exer. 6.8 for the gauge integral).

23*. Let $I = [a, b]$. Show that the inclusion map $\mathcal{M}(I) \to \mathcal{HK}(I)$ is continuous with respect to the norm $\| \|_1$ on $\mathcal{M}(I)$ and the Alexiewicz norm on $\mathcal{HK}(I)$.

24. Show that the product of McShane integrable functions need not be McShane integrable. Hint: t^p.
25. Show that the assumption in Theorem 29 that g is bounded and McShane integrable over I can be replaced by the assumption that g is bounded and

McShane integrable over every bounded subinterval of I. [Hint: Consider gC_{I_k} where $I_k = [-k, k] \cap I$.]

26. Give an example where Exercise 25 applies but Theorem 29 does not.
27. Show that $\mathcal{M}(I)$ is separable under $\| \|_1$.
28. Let $f : \mathbb{R} \to \mathbb{R}$ be McShane integrals. For $k \in \mathbb{N}$ let

$$f_k(t) = \begin{cases} f(t) & |f(t)| \le k \\ k & f(t) > k \\ -k & f(t) < -k \end{cases}$$

be the truncation of f at k. Show that each f_k is McShane integrable and $\int_{\mathbb{R}} f_k \to \int_{\mathbb{R}} f$. Show that an analogous result is false for the gauge integral. [See [B] for a discussion of this truncation for the gauge integral; another truncation for the gauge integral is considered in [LPY] § 18.]

Chapter 10

McShane Integrability is Equivalent to Absolute Henstock-Kurzweil Integrability

It is clear from the definitions (1.2, 4.2, 9.1) that if a function is McShane integrable, then it is Henstock-Kurzweil (gauge) integrable and both integrals have the same value. Example 2.12 furnishes an example of a function which is Henstock-Kurzweil integrable but is not McShane integrable (see also Exercises 3.9, 4.3). In this chapter we establish the basic connection between the two integrals. In particular, we give an elementary proof of the fact that a function is McShane integrable over a closed, bounded interval if and only if it is absolutely Henstock-Kurzweil integrable.

We begin by fixing the notation which will be employed. Let $I = [a, b]$, $-\infty < a < b < \infty$, and $f : I \to \mathbb{R}$. It is convenient in this chapter to consider a gauge to be a positive function $\delta : I \to (0, \infty)$ as in Chapter 1 (recall that such a function generates a gauge γ by setting $\gamma(t) = (t - \delta(t), t + \delta(t))$ so a tagged partition $\mathcal{D} = \{(t_i, I_i) : 1 \le i \le m\}$ is δ-fine if and only if $t_i \in I_i \subset (t_i - \delta(t_i), t_i + \delta(t_i))$ for $1 \le i \le m$; in this case we write $\mathcal{D} << \delta$ as before).

Let δ be a gauge on I. If $a < x \le b$, we can consider δ to be a gauge in $[a, x]$; let $\pi_\delta[a, x]$ be all δ-fine tagged partitions of $[a, x]$. We define 2 functions M_δ, m_δ on I by

$$M_\delta(a) = m_\delta(a) = 0\,,$$

$$M_\delta(x) = \sup\{S(f, \mathcal{D}) : \mathcal{D} \in \pi_\delta[a, x]\}\,,$$

$$m_\delta(x) = \inf\{S(f, \mathcal{D}) : \mathcal{D} \in \pi_\delta[a, x]\}\,,$$

for $a < x \leq b$; the function $M_\delta(m_\delta)$ is called a *major* (*minor*) function for f [see the Notes/Remarks section, Lemma 1 and Exercise 2 which give a motivation for these definitions and also the method of proof]. If f is Henstock-Kurzweil (gauge) integrable over I, we write $H\int_a^b f$ for its integral, and if f is McShane integrable over I, we write $M\int_a^b f$ for its integral.

Lemma 1. (i) *If $x-\delta(x) < u \leq x \leq v < x+\delta(x)$, then $M_\delta(v) - M_\delta(u) \geq f(x)(v-u)$.*

(ii) *If $x-\delta(x) < u \leq x \leq v < x+\delta(x)$, then $m_\delta(v) - m_\delta(u) \leq f(x)(v-u)$.*

(iii) *$M_\delta - m_\delta$ is non-negative and increasing on I.*

(iv) *If $f \geq 0$, then both M_δ and m_δ are non-negative and increasing.*

(v) *If f is Henstock-Kurzweil integrable over I and if $\varepsilon > 0$ and δ is a gauge on I such that*

$$\left| S(f,\mathcal{D}) - H\int_a^b f \right| < \varepsilon$$

whenever $\mathcal{D} \in \pi_\delta[a,b]$, then $0 \leq M_\delta(b) - m_\delta(b) \leq 2\varepsilon$.

Proof: (i): Let $\mathcal{D} \in \pi_\delta[a,u]$. Then $\mathcal{D} \cup \{(x,[u,v])\} \in \pi_\delta[a,v]$ so

$$M_\delta(v) \geq S(f,\mathcal{D}) + f(x)(v-u).$$

Hence,

$$M_\delta(v) \geq M_\delta(u) + f(x)(v-u)$$

and (i) follows.

(ii) is similar to (i).

(iii) Let $\varepsilon > 0$ and $a \leq u < v \leq b$. There exists $\mathcal{D}, \mathcal{D}' \in \pi_\delta[a,u]$ such that

$$M_\delta(u) - m_\delta(u) \leq S(f,\mathcal{D}) - S(f,\mathcal{D}') + \varepsilon.$$

Let $\mathcal{F} \in \pi_\delta[u,v]$. Then $\mathcal{E} = \mathcal{D} \cup \mathcal{F}, \mathcal{E}' = \mathcal{D}' \cup \mathcal{F} \in \pi_\delta[a,v]$ so

$$\begin{aligned} M_\delta(u) - m_\delta(u) &\leq S(f,\mathcal{D}) - S(f,\mathcal{D}') + \varepsilon = S(f,\mathcal{E}) - S(f,\mathcal{E}') + \varepsilon \\ &\leq M_\delta(v) - m_\delta(v) + \varepsilon, \end{aligned}$$

and therefore,

$$M_\delta(u) - m_\delta(u) \leq M_\delta(v) - m_\delta(v)$$

so $M_\delta - m_\delta$ is increasing. Clearly, $M_\delta - m_\delta$ is non-negative.

(iv) is similar to (iii).

For (v), if $\mathcal{D}, \mathcal{E} \in \pi_\delta[a,b]$, then $|S(f,\mathcal{D}) - S(f,\mathcal{E})| < 2\varepsilon$ so $M_\delta(b) - m_\delta(b) \leq 2\varepsilon$.

The proof below makes repeated use of Fatou's Lemma. We use the version of Fatou's Lemma given in Remark 9.38 so it needs to be realized that there are functions involved which may take on infinite values on a null set.

Theorem 2. *f is McShane integrable over I if and only if f is absolutely Henstock-Kurzweil integrable over I.*

Proof: As noted earlier since the McShane integral is an absolute integral, if f is McShane integrable, then f is absolutely Henstock-Kurzweil integrable.

For the converse, it suffices to show that if f is non-negative and Henstock-Kurzweil integrable, then f is McShane integrable. For convenience, extend f to $[a,\infty)$ by setting $f(t) = 0$ for $b < t < \infty$. Let $\varepsilon > 0$ and pick a gauge δ on I such that if $\mathcal{D}$ is a δ-fine tagged partition of I, then $|S(f,\mathcal{D}) - H\int_a^b f| < \varepsilon$. Extend M_δ and m_δ by setting $M_\delta(t) = M_\delta(b)$, $m_\delta(t) = m_\delta(b)$ for $t > b$. Define

$$H_n(t) = n(M_\delta(t+1/n) - M_\delta(t)) \text{ and } h_n(t) = n(m_\delta(t+1/n) - m_\delta(t)).$$

Since M_δ and m_δ are increasing (Lemma 1 (iv)), H_n and h_n are McShane integrable (Exercise 9.2). Set $H = \underline{\lim} H_n$, $h = \overline{\lim} h_n$.

We claim that

$$(1)\quad 0 \leq M\int_a^b H_n \leq M_\delta(b).$$

Indeed, we have

$$0 \leq M\int_a^b H_n = n\left(M\int_b^{b+1/n} M_\delta - M\int_a^{a+1/n} M_\delta\right)$$
$$\leq n\left(M\int_b^{b+1/n} M_\delta\right) = M_\delta(b)$$

(Exercise 9.3 and Lemma 1. (iv)).

Fatou's Lemma (Remark 9.38) implies that H is McShane integrable and $M\int_a^b H \leq M_\delta(b)$.

We next claim that h is McShane integrable. For this suppose t is fixed and $n > 1/\delta(t)$. Then, by Lemma 1 (i) and (ii), for large n,

$$f(t) \leq n(M_\delta(t+1/n) - M_\delta(t)) \text{ and } f(t) \geq n(m_\delta(t+1/n) - m_\delta(t))$$

so $f(t) \le H(t)$, $f(t) \ge h_n(t)$. Therefore, $\underline{\lim}(H - h_n)^+ = H - h$. As in (1), $0 \le M\int_a^b h_n \le m_\delta(b)$ so Fatou's Lemma implies that $H - h$ is McShane integrable and, hence, h is McShane integrable.

Note that from the computations above we have

(2) $h(t) \le f(t) \le H(t)$.

Finally, we claim that

(3) $0 \le M\int_a^b (H - h) \le 2\varepsilon$.

As in (1), it follows that

$$0 \le M\int_a^b (H_n - h_n) \le M_\delta(b) - m_\delta(b)\,.$$

Since $H - h \le \underline{\lim}(H - h_n)$, Fatou's Lemma and Lemma 1 (v) give

(4) $M\int_a^b (H - h) \le M\int_a^b \underline{\lim}(H_n - h_n) \le \underline{\lim} M\int_a^b (H_n - h_n) \le M\delta(b) - m\delta(b) \le 2\varepsilon$.

It follows from (2) and (3) that f is McShane integrable (9.3 (iv)).

Of course, it is also the case that $H\int_a^b f = M\int_a^b f$.

Notes/Remarks

The proof of Theorem 2 is due to Vyborny ([V]). There were several earlier proofs of the result, but all of these proofs used Lebesgue measure/integration — a concept not present in the definitions of either integral (see, for example, [ML] § 8.3, [LPY] 16.10, [LW], [Pf1] 6.3.4, [C]). Another elementary proof is given in [CWT].

The definitions of the major and minor functions are taken from [BV] but essentially appeared in [K1].

The appearance of major/minor functions in the proof of the equivalence of the McShane and absolute Henstock-Kurzweil integrals is not surprising. It is known that the Henstock-Kurzweil integral is equivalent to the Perron integral ([Go1], [LPY] § 8) and major/minor functions are employed in the definition of the Perron integral [the definition of major/minor functions for the Perron integral is different from the definitions used above, but the functions M_δ, m_δ

are major/minor functions in the Perron sense; see the definition of the Perron integral below and Exercise 2].

In order to appreciate the proof of Theorem 2 above we give a brief description of the Perron integral. A function $H : I \to \mathbb{R}$ is a *major function* of f if $H(a) = 0$ and $\underline{D}H(t) \geq f(t)$ for $t \in I; h : I \to \mathbb{R}$ is a *minor function* for f if $h(a) = 0$ and $\overline{D}f(t) \leq f(t)$ for $t \in I$ [here $\overline{D}f(\underline{D}f)$ is the upper (lower) derivative of f].

Definition 3. *f is Perron integrable over I if f has both major and minor functions and*

$$-\infty < \inf\{H(b) : H a\ major\ function\ for\ f\}$$
$$= \sup\{h(b) : h a\ minor\ function\ for\ f\} < \infty\,.$$

The common value A in Definition 3 is called the Perron integral of f over I and is denoted by $P\int_a^b f = A$.

We have the following criterion for Perron integrability.

Theorem 4. *f is Perron integrable over I if and only if for every $\varepsilon > 0$ there exist major and minor functions, H and h, for f such that $0 \leq H(b) - h(b) < \varepsilon$.*

Note that Eq. (4) is just the sufficient condition of Theorem 4.

For the basic properties of the Perron integral, including Theorem 4, see [N2] Chapter XVI or [Go1].

Exercises

1. Extend Theorem 2 to unbounded intervals.
2. Show that the functions M_δ, m_δ are major, minor functions for f in the sense of Definition 3 for the Perron integral.

Appendix 1: The Riemann Integral

In this appendix we repeat Riemann's definition of the Riemann integral and give references where the interested reader can find expositions of the integral.

Let $I = [a, b]$ and $f : I \to \mathbb{R}$.

Definition 1. *The function f is Riemann integrable over I if there exists $A \in \mathbb{R}$ such that for every $\varepsilon > 0$ there is a $\delta > 0$ such that if $\mathcal{P} = \{a = x_0 < x_1 < \cdots < x_n = b\}$ is a partition of I with $x_i - x_{i-1} < \delta$ for $1 \le i \le n$ and $x_{i-1} \le t_i \le x_i$, then*

$$\left| \sum_{i=1}^{n} f(t_i)(x_i - x_{i-1})v - A \right| < \varepsilon .$$

The term $\sum_{i=1}^{n} f(t_i)(x_i - x_{i-1})$ is called a *Riemann sum* of f with respect to $\mathcal{P}$ and the choice of points t_i. The number A is the Riemann integral of f over I and is usually denoted by $\int_I f$, $\int_a^b f$, $\int_a^b f(t)dt$ or $\int_I f(t)dt$.

For an exposition of the Riemann integral using Definition 1, see Hoffman, [Ho] Chapter 4. There is another approach to the Riemann integral due to Darboux that is more popular in the current introductory texts on real analysis; this approach, based on upper and lower integrals, is also discussed in Hoffman and the 2 approaches are shown to be equivalent (see also [Ru]).

Exercises

1. If f is Riemann integrable over $[a, b]$, show that f is bounded on $[a, b]$.
2. Define $f : [a, b] \to R$ by $f(t) = 0$ if t is irrational and $f(t) = 1$ if t is rational [f is often called the Dirichlet function]. Show that f is not Riemann integrable.

Appendix 2: Functions of Bounded Variations

For the reader who may not be familiar with the basic properties of functions of bounded variation, we record them in this appendix.

Let $f : [a, b] \to \mathbb{R}$. If $\pi = \{a = x_0 < x_1 < \cdots < x_n = b\}$ is a partition of $[a, b]$, the variation of f over π is

$$var(f : \pi) = \sum_{i=0}^{n-1} |f(x_{i+1}) - f(x_i)|,$$

and the variation of f over $[a, b]$ is

$$Var(f : [a, b]) = \sup var(f : \pi),$$

where the supremum is taken over all possible partitions, π, of $[a, b]$. If

$$Var(f : [a, b]) < \infty,$$

f is said to have bounded variation; the class of all such functions is denoted by $BV[a, b]$. The variation measures the amount the function oscillates in $[a, b]$.

As the example below illustrates, even a continuous function can fail to belong to $BV[a, b]$.

Example 1. *Let $f(t) = t\sin(1/t)$ for $0 < t \le 1$ and $f(0) = 0$. Set $x_n = 1/(n+1/2)\pi$. Then $f(x_n) = 1/(n + 1/2)\pi$ if n is even, and $f(x_n) = -1/(n + 1/2)\pi$ if n is odd. If π_n is the partition $\{0 < x_n < x_{n-1} < \cdots < x_1 < 1\}$, then*

$$\sum_{i=1}^{n-1} |f(x_i) - f(x_{i-1})| \ge \frac{2}{\pi} \sum_{i=1}^{n-1} 1/(i+1)$$

so $Var(f : [0, 1]) = \infty$.

Proposition 2. *If $f \in BV[a,b]$, then f is bounded on $[a,b]$.*

Proof: Let $x \in (a,b)$. Then

$$|f(x) - f(a)| + |f(b) - f(x)| \le Var(f : [a,b])$$

so

$$2|f(x)| \le |f(a)| + |f(b)| + Var(f : [a,b])$$

and f is bounded.

We consider properties of $Var(f : I)$ as a function of the interval I. First, as a consequence of the triangle inequality, we have

Lemma 3. *Let $f : [a,b] \to \mathbb{R}$. If π and π' are partitions of $[a,b]$ with $\pi \subset \pi'$, then $var(f : \pi) \le var(f : \pi')$.*

Proposition 4. *Let $f : [a,b] \to \mathbb{R}$ and $a < c < b$. Then*

$$Var(f : [a,b]) = Var(f : [a,c]) + Var(f : [c,b]) .$$

Proof: Let π be a partition of $[a,b]$ and π' the partition obtained by adding the point c to π. Let π_1 and π_2 be the partitions of $[a,c]$ and $[c,b]$, respectively, induced by π'. Then by Lemma 3,

$$var(f : \pi) \le var(f : \pi') = var(f : \pi_1) + var(f : \pi_2) \le Var(f : [a,c]) + Var(f : [c,b])$$

so

$$Var(f : [a,b]) \le Var(f : [a,c]) + Var(f : [c,b]) .$$

If π_1 and π_2 are partitions of $[a,c]$ and $[c,b]$, respectively, then $\pi = \pi_1 \cup \pi_2$ is a partition of $[a,b]$ so

$$var(f : \pi) = var(f : \pi_1) + var(f : \pi_2) \le Var(f : [a,b]) .$$

Hence,

$$Var(f : [a,c]) + Var(f : [c,b]) \le Var(f : [a,b]) .$$

Next, we consider how the variation, $Var(f : I)$, depends on the function f.

Proposition 5. *Let $f, g \in BV[a,b]$. Then*

(i) $f+g \in BV[a,b]$ *with* $Var(f+g:[a,b]) \leq Var(f:[a,b]) + Var(g:[a,b])$,

(ii) *for* $t \in \mathbb{R}, tf \in BV[a,b]$ *with* $Var(tf:[a,b]) = |t| Var(f:[a,b])$.

Proof: (i) follows from the triangle inequality and (ii) is clear.

Thus, $BV[a,b]$ is a vector space under the usual pointwise addition and scalar multiplication of functions.

Let $f \in BV[a,b]$. We define the total variation of f by $V_f(t) = Var(f: [a,t])$ if $a < t \leq b$ and $V_f(a) = 0$.

Proposition 6. *V_f and $V_f - f$ are increasing on $[a,b]$.*

Proof: V_f is increasing by Proposition 4.

Let $a \leq x < y \leq b$ and $g = V_f - f$. Then

$$g(y) = V_f(x) + Var(f:[x,y]) - f(y)$$

implies

$$\begin{aligned} g(y) - g(x) &= V_f(x) - f(y) + Var(f:[x,y]) - V_f(x) + f(x) \\ &= Var(f:[x,y]) - (f(y) - f(x)) \geq 0\,. \end{aligned}$$

Proposition 7. *If $f \in BV[a,b]$ is (right, left) continuous at x, then V_f is (right, left) continuous at x.*

Proof: Let $\epsilon > 0$. Suppose f is right continuous at $x < b$. There is a partition π of $[x,b]$ such that $var(f:\pi) > Var(f:[x,b]) - \epsilon$. Since f is right continuous at x, we may add a point x_1 to π to obtain a partition

$$\pi' = \{x < x_1 < \cdots < x_n\}$$

of $[x,b]$ such that $|f(x) - f(x_1)| < \epsilon$. Then

$$\begin{aligned} \epsilon + var(f,\pi') &= \epsilon + |f(x_0) + f(x_1)| + \sum_{i=1}^{n-1} |f(x_{i+1}) - f(x_i)| \\ &< 2\epsilon + Var(f:[x_1,b]) \end{aligned}$$

so

$$Var(f:[x,b]) < var(f:\pi) + \epsilon \leq var(f:\pi') + \epsilon < 2\epsilon + Var(f:[x_1,b])\,.$$

Thus, $0 \le V_f(x_1) - V_f(x) < 2\epsilon$, and since $V_f \uparrow, \lim_{y \to x^+} V_f(y) = V_f(x)$, i.e., V_f is right continuous.

The statement about left continuity is similar.

Since $f = V_f - (V_f - f)$, Propositions 6 and 7 along with Exercise 1 and Proposition 5 give the following characterization of functions of bounded variation.

Theorem 8. *Let $f : [a,b] \to \mathbb{R}$. Then $f \in BV[a,b]$ if and only if $f = g - h$, where $g, h \uparrow$. If f is (right, left) continuous, then g and h can be chosen to be (right, left) continuous.*

We now extend the notion of bounded variation to functions defined on unbounded intervals. Let $f : \mathbb{R} \to \mathbb{R}$.

Definition 9.

$$Var(f, \mathbb{R}) = \sup\{Var(f : [a,b]) : -\infty < a < b < \infty\},$$

$$Var(f : [a,\infty)) = \sup\{Var(f : [a,b]) : a < b < \infty\},$$

$$Var(f : (-\infty, b]) = \sup\{Var(f : [a,b]) : -\infty < a < b\}.$$

We say that f has *bounded variation* over the interval I if $Var(f : I) < \infty$; we denote the space of all such functions by $BV(I)$.

The following properties are easily derived.

Proposition 10.

(i)
$$Var(f : \mathbb{R}) = \lim_{a \to \infty} Var(f : [-a,a]),$$

$$Var(f : [a,\infty)) = \lim_{b \to \infty} Var(f : [a,b]),$$

$$Var(f : (-\infty,b]) = \lim_{a \to -\infty} Var(f : [a,b]).$$

(ii) *For $a < b$,* $Var(f : \mathbb{R}) = Var(f : (-\infty,a]) + Var(f : [a,\infty))$,

$$Var(f : (-\infty,b]) = Var(f : (-\infty,a]) + Var(f : [a,b]),$$

$$Var(f : [a,\infty)) = Var(f : [a,b]) + Var(f : [b,\infty)).$$

We have the following further properties of the variation function.

Proposition 11. *Assume* $Var(f, \mathbb{R}) < \infty$. *Then*

(i) f *is bounded and*
(ii) $\lim_{x\to-\infty} Var(f:(-\infty,x]) = 0, \lim_{x\to\infty} Var(f:[x,\infty)) = 0$.

Proof: (i): Fix $a \in \mathbb{R}$. For $x > a$,

$$|f(x) - f(a)| \le Var(f:[a,x]) \le Var(f:\mathbb{R});$$

similarly, for $x < a$,

$$|f(a) - f(x)| \le Var(f:[x,a]) \le Var(f:\mathbb{R}).$$

Hence,

$$|f(x)| \le |f(a)| + Var(f:\mathbb{R}).$$

(ii): Let $\varepsilon > 0$. There exists a such that $Var(f:\mathbb{R}) < Var(f:(-\infty,a])+\varepsilon$. Since

$$Var(f:\mathbb{R}) = Var(f:(-\infty,a]) + Var(f:[a,\infty)), Var(f:[a,\infty)) < \varepsilon$$

so $\lim_{x\to\infty} Var(f:[x,\infty)) = 0$. The other statement is similar.

It is easily seen that the analogue of Proposition 5 holds for unbounded intervals. If $f \in BV(\mathbb{R})$, we define the total variation of f by $V_f(t) = Var(f:(-\infty,t])$. The analogues of Propositions 6 and 7 then hold for the total variation function and, thus, Theorem 8 likewise holds for functions of bounded variation on $\mathbb{R}$.

Exercises

1. If f is increasing on $[a,b]$, show $Var(f:[a,b]) = f(b) - f(a)$. If f is increasing on $\mathbb{R}$, give necessary and sufficient conditions for f to be of bounded variation on $\mathbb{R}$.
2. Give necessary and sufficient conditions for f to satisfy $Var(f:[a,b]) = 0$. Repeat for $Var(f:\mathbb{R}) = 0$.
3. Let I be a closed interval in $\mathbb{R}$ and let $f, g \in BV(I)$. Show that $fg, |f|, f \vee g, f \wedge g$ belong to $BV(I)$.
4. Prove Proposition 10.
5. Let $f \in BV(\mathbb{R})$. Show that $\lim_{x\to\pm\infty} f(x)$ exists.

Appendix 3: Differentiating Indefinite Integrals

In this appendix we consider the other half of the Fundamental Theorem of Calculus for the gauge integral, differentiating indefinite integrals.

Let $I = [a,b], -\infty < a < b < \infty$. If $f \in \mathcal{HK}(I)$, the space of gauge integrable functions on I, and $F(t) = \int_a^t f, a \le t \le b$, we show that F is differentiable a.e. with $F' = f$ a.e.

For the proof we require a covering theorem of Vitali. If $E \subset I$, a family $\mathcal{J}$ of closed subintervals is a *Vitali covering* of E if for every $t \in E$ and $\delta > 0$ there exists $J \in \mathcal{J}$ such that $t \in J$ and $\ell(J) < \delta$.

Let $\lambda\,(\lambda^*)$ denote Lebesgue (outer) measure on $\mathbb{R}$.

Theorem 1 (*Vitali Covering Theorem*). *Let $E \subset I$ and let $\mathcal{J}$ be a Vitali covering of E. If $\varepsilon > 0$, there is a finite, pairwise disjoint subset $\{J_k : 1 \le k \le m\}$ of $\mathcal{J}$ such that $\lambda^*(E) \le \sum_{k=1}^m \ell(J_k) + \varepsilon$.*

For a proof of this (and further) result, see [Ro], 5.1.1, [N 1].

Theorem 2. *Let $f \in \mathcal{HK}(I)$ and let $F(t) = \int_a^t f, a \le t \le b$, be the indefinite integral of f. Then F is differentiable a.e. in I with $F' = f$ a.e.*

Proof: Let E be the points in I where either F fails to have a right hand derivative or the right hand derivative exists but does not equal f. We show that E is a null set [has Lebesgue measure 0]. Let $\varepsilon > 0$. There exists a gauge γ on I such that if $\mathcal{D} = \{(t_k, [x_{k-1}, x_k]) : 1 \le k \le m\} << \gamma$, then

$$\left| \sum_{k=1}^{n} \{F(x_k) - F(x_{k-1}) - f(t_k)(x_k - x_{k-1})\} \right| < \varepsilon .$$

We may assume that $\gamma(t) = (t - \delta(t), t + \delta(t))$ with $\delta(t) > 0$ for $t \in I$. For every $t \in E$ there exists $\eta(t) > 0$ such that for every $\delta < \delta(t)$ there is a point

$s \in I$ with $0 < s - t < \delta$ and

$$|F(t) - F(s) - f(t)(s-t)| \geq \eta(t)(s-t)\,.$$

For $n \in \mathbb{N}$, set $E_n = \{t \in E : \eta(t) \geq 1/n\}$. To show $\lambda(E) = 0$ it suffices to show that $\lambda(E_n) = 0$ for every n.

Fix n and let $\mathcal{J}$ be the family of all closed intervals $[t,s]$ with $t \in E_n$ and s as above. Then $\mathcal{J}$ is a a Vitali cover of E_n. By the Vitali covering theorem, there exists $\{[t_k, s_k] : 1 \leq k \leq m\} \subset \mathcal{J}$ pairwise disjoint with $\lambda^*(E_n) \leq \sum_{k=1}^{m}(s_k - t_k) + \varepsilon$. Now $\{(t_k, [t_k, s_k]) : 1 \leq k \leq m\}$ is a γ-fine partial tagged division of I so by Lemma 9.3 we obtain

$$\lambda^*(E_n) \leq \sum_{k=1}^{m} |F(t_k) - F(s_k) - f(t_k)(s_k - t_k)|/\eta(t_k) + \varepsilon \leq 2n\varepsilon + \varepsilon\,.$$

Hence, $\lambda^*(E_n) = 0$ as desired.

We can treat the points in I where F has a left-hand derivative in a similar manner thus establishing the result.

A consequence of Theorem 2 is that any gauge integrable function is (Lebesgue) measurable.

Theorem 3. *If $f \in \mathcal{HK}(I)$, then f is (Lebesgue) measurable.*

Proof: F is continuous by Corollary 3.2 and therefore measurable. By Theorem 2

$$\lim_{k\to\infty} \frac{F(t+1/k) - F(t)}{1/k} = f(t)$$

for all t except those in a null set so f is measurable.

Notes/Remarks

For other proofs of the measurability see [LPY] 5.10, [LPY1] or [Pf1] 6.3.3.

Appendix 4: Equivalence of Lebesgue and McShane Integrals

In this section we show the equivalence of the Lebesgue and McShane integrals over closed, bounded intervals.

Again λ denotes Lebesgue measure.

Theorem 1. *Let $f : [a, b] \to \mathbb{R}$ be Lebesgue integrable over $[a, b]$. Then f is McShane integrable over $[a, b]$ and the values of the two integrals are equal.*

Proof: We may assume that f is non-negative ([Sw1] 3.2.11). All integrals in the proof below will be Lebesgue integrals. Let $\varepsilon > 0$. Choose $\delta > 0$ such that $\lambda(A) < \delta$ implies $\int_A f < \varepsilon$ ([Sw1] 3.2.17).

Let $\alpha = \min\{1, \varepsilon/(\delta + b - a)\}$. For $n \in \mathbb{N}$, let

$$E_n = \{t : (n-1)\alpha \le f(t) < n\alpha\}$$

so $\{E_n : n \in \mathbb{N}\}$ are pairwise disjoint with $[a, b] = \bigcup_{n=1}^{\infty} E_n$. For each n pick G_n open such that $G_n \supset E_n$ and $\lambda(G_n \backslash E_n) < \delta/2^n(n+1)$. Define a gauge γ on $[a, b]$ by choosing $\delta(t)$ to be an open interval containing t and such that $\delta(t) \subset G_n$ whenever $t \in E_n$.

Now suppose that $\mathcal{D} = \{(t_i, I_i) : 1 \le i \le m\}$ is a γ-fine free tagged partition of $[a, b]$. We show

$$\left| S(f, \mathcal{D}) - \int_a^b f \right| < 3\varepsilon ,$$

completing the proof. For each $i = 1, \dots, m$, let n_i be such that $t_i \in E_{n_i}$. Then

$$(1)\qquad \left|S(f,\mathcal{D})-\int_a^b f\right| \le \sum_{i=1}^m \int_{I_i} |f(t_i)-f(t)|dt$$
$$\le \sum_{i=1}^m \int_{I_i\cap E_{n_i}} |f(t_i)-f(t)|dt$$
$$+\sum_{i=1}^m \int_{I_i\backslash E_{n_i}} f(t_i)dt + \sum_{i=1}^m \int_{I_i\backslash E_{n_i}} f(t)dt$$
$$= R_1+R_2+R_3\,,$$

with obvious definitions of R_i.

To estimate R_1, note that if $t_i, t \in E_{n_i}$, then both $f(t_i)$ and $f(t)$ belong to the interval $[(n_i-1)\alpha, n_i\alpha)$ so

$$R_1 \le \sum_{i=1}^m \alpha\lambda(I_i\cap E_{n_i}) \le \alpha(b-a) < \varepsilon\,.$$

Since $t_i \in E_{n_i}$ and $I_i \subset G_{n_i}$ ($\mathcal{D}$ is γ-fine),

$$R_2 = \sum_{n=1}^\infty \sum_{\substack{i\\ n_i=n}} \int_{I_i\backslash E_{n_i}} f(t_i)dt \le \sum_{n=1}^\infty \sum_{\substack{i\\ n_i=n}} (n+1)\alpha\lambda(I_i\backslash E_{n_i})$$

$$\le \sum_{n=1}^\infty (n+1)\alpha\lambda(G_n\backslash E_n) \le \sum_{n=1}^\infty (n+1)\alpha\frac{\delta}{2^n(n+1)} \le \sum_{n=1}^\infty \frac{\varepsilon}{2^n} = \varepsilon\,.$$

For R_3 put $A=\bigcup_{i=1}^m (I_i\backslash E_{n_i})$. Since $I_i \subset G_{n_i}$,

$$\lambda(A) = \sum_{i=1}^m \lambda(I_i\backslash E_{n_i}) = \sum_{n=1}^\infty \sum_{\substack{i\\ n_i=n}} \lambda(I_i\backslash E_{n_i})$$

$$\le \sum_{n=1}^\infty \lambda(G_n\backslash E_n) \le \sum_{n=1}^\infty \frac{\delta}{2^n(n+1)} < \delta\,.$$

Thus, $R_3 \le \int_A f < \varepsilon$.

From (1), we have $|S(f,\mathcal{D})-\int_a^b f| < 3\varepsilon$ as desired.

Theorem 2. *If $f:[a,b]\to\mathbb{R}$ is McShane integrable, then f is Lebesgue integrable over $[a,b]$ and the two integrals are equal.*

Proof: All integrals in the proof below will be McShane integrals. Let $F(t) = \int_a^t f$ be the indefinite integral of f. Since f is McShane integrable, f is Henstock-Kurzweil integrable so F is differentiable a.e. in $[a, b]$ with $F' = f$ a.e. in $[a, b]$ (Appendix 3.2). If we show that $F \in BV[a, b]$, then f is Lebesgue integrable ([Sw1] 4.2.5 and A.1.8).

Let γ be a gauge on $[a, b]$ such that $|S(f, \mathcal{D}) - \int_a^b f| < 1$ whenever $\mathcal{D}$ is a γ-fine free tagged partition of $[a, b]$. Fix such a $\mathcal{D} = \{(t_i, I_i) : 1 \le i \le m\}$. It sufficies to show that F is of bounded variation on each I_i. Fix i, $1 \le i \le m$. Let $x_0 < x_1 < \cdots < x_n$ be a partition of I_i. Then $\mathcal{E} = \{(t_i, [x_{j-1}, x_j]) : 1 \le j \le n\}$ is a γ-fine free tagged partition of I_i. By Lemma 9.3,

$$\sum_{j=1}^{n} \left| f(t_i)(x_j - x_{j-1}) - \int_{x_{j-1}}^{x_j} f \right| \le 2$$

which implies $\sum_{j=1}^{n} |F(x_j) - F(x_{j-1})| \le 2 + |f(t_i)|\ell(I_i)$. Hence, F is of bounded variation on I_i.

The 2 integrals are equal by Theorem 1.

Remark 3. *The construction of the free tagged partition $\mathcal{E}$ in the proof above shows the utility of free tags.*

Notes/Remarks

The proof of Theorem 1 is that of Davies and Schuss ([DSc]). The proof of Theorem 2 is taken from Mendoza ([Me]).

Other proofs of the equivalence of the Lebesgue and McShane integrals can be found in [Lin], [Go1] 10.11, [Ku].

Exercise

1. Extend Theorems 1 and 2 to unbounded intervals.

Appendix 5: Change of Variable in Multiple Integrals

In this appendix we establish several versions of the change of variable theorem for multiple integrals. The first version is for continuous functions and requires only basic results for the gauge integral. A general version is then established which requires fairly extensive use of results for the Lebesgue integral. We begin by considering the case of linear transformations on $\mathbb{R}^n$. First, consider the following special linear transformations on $\mathbb{R}^n$.

(i) *For* $t \neq 0$, *define* $d_t : \mathbb{R}^n \to \mathbb{R}^n$ *by* $d_t(x_1, \ldots, x_n) = (tx_1, \ldots, x_n)$.
(ii) *Define* $a : \mathbb{R}^n \to \mathbb{R}^n$ *by* $a(x_1, \ldots, x_n) = (x_1 + x_2, x_2, \ldots, x_n)$.
(iii) *For* $1 \leq i < j \leq n$, *define* $s_{ij} : \mathbb{R}^n \to \mathbb{R}^n$ *by*

$$s_{ij}(x_1, \ldots, x_i, \ldots, x_j, \ldots, x_n) = (x_1, \ldots, x_j, \ldots, x_i, \ldots, x_n).$$

If A is an $n \times n$ matrix, its determinant is denoted by $\det A$.

Lemma 1. *If A is any of the linear transformations in* (i), (ii) *or* (iii) *and $f : \mathbb{R}^n \to \mathbb{R}$ is absolutely integrable, then $f \circ A$ is absolutely integrable with* $\int_{\mathbb{R}^n} f(y)dy = \int_{\mathbb{R}^n} f(Ax)|\det A|dx$.

Proof: Each of the cases in (i), (ii) or (iii) follows easily from Fubini's Theorem 8.13 and the linear change of variable formula in $\mathbb{R}^1$. (Exercise 1.) For example, if $t > 0$, then

$$\int_{\mathbb{R}^n} f(d_t x)dx = \int_{-\infty}^{\infty} \cdots \int_{-\infty}^{\infty} f(tx_1, \ldots x_n)\, dx_1 \cdots dx_n$$

$$= \int_{-\infty}^{\infty} \cdots \int_{-\infty}^{\infty} \frac{1}{t} f(x_1, \ldots, x_n)dx_1 \cdots dx_n$$

$$= (1/|\det d_t|) \int_{\mathbb{R}^n} f(x)dx\,.$$

Every non-singular linear transformation $A : \mathbb{R}^n \to \mathbb{R}^n$ is a product of linear transformations of the form in (i), (ii) and (iii) ([HK] 1.6.12) so we obtain

Theorem 2. *If $A : \mathbb{R}^n \to \mathbb{R}^n$ is linear and $f : \mathbb{R}^n \to \mathbb{R}$ is absolutely integrable, then $f \circ A$ is absolutely integrable with*

(1) $\displaystyle\int_{\mathbb{R}^n} f(y)dy = \int_{\mathbb{R}^n} f(Ax)|\det A|dx.$

We denote Lebesgue measure on $\mathbb{R}^n$ by λ. From Theorem 2 we have

Corollary 3. *If $B \subset \mathbb{R}^n$ is a bounded Borel set and $A : \mathbb{R}^n \to \mathbb{R}^n$ is linear and nonsingular, then*

(2) $\lambda(A^{-1}B) = (\det A^{-1})\lambda(B)$.

Proof: Put $f = C_B$ in (1) to obtain

$$\int_{\mathbb{R}^n} f = \lambda(B) = \int_{\mathbb{R}^n} f(Ax)|\det A|dx = \int_{\mathbb{R}^n} C_{A^{-1}B}(x)|\det A|dx$$
$$= \lambda(A^{-1}B)|\det A|\,.$$

We now consider the case of nonlinear transformations on $\mathbb{R}^n$. We assume the reader is familiar with the basic properties of differentiable functions from $\mathbb{R}^n \to \mathbb{R}^n$ (for example, see [DS] §29, [Ru] §9). We begin by setting down the basic assumptions and notations. Let D_1 and D_2 be open subsets of $\mathbb{R}^n$ with $h = (h_1, \ldots, h_n) : D_1 \to \mathbb{R}^n$ a continuously differentiable function from D_1 onto D_2 with a continuously differentiable inverse $g : D_2 \to D_1$. If $f : D_1 \to \mathbb{R}$, we denote the jth partial derivative of f by $\partial_j f$, and we denote the differential of h by Dh. Thus, Dh is the linear transformation on $\mathbb{R}^n$ whose (Jacobian) matrix is given by $Dh = [\partial_j h_i]$ and by the Chain Rule Dg is the inverse of Dh.

For the case of a general transformation h we consider dividing the region D_1 into cubes. If C is any cube inside D_1 and $x \in C$, then $h(x)$ is approximated by $Dh(x)$, and if h were linear, then by Corollary 3 $\lambda(h(C)) = |\det Dh(x)|\lambda(C)$. Thus, it is reasonable to assume that in the nonlinear case $\lambda(h(C))$ is approximated by $|\det Dh(x)|\lambda(C)$. We establish an upper bound for the approximation in Lemma 4 and use a device of J. Schwartz ([Sz]) to establish the transformation formula

(3) $\displaystyle\int_{D_2} f(y)dy = \int_{D_1} f(h(x))|\det Dh(x)|dx$

for continuous functions.

In what follows it is convenient to use the sup-norm,

$$\|x\| = \|(x_1, \dots, x_n)\| = \max\{|x_i| : 1 \le i \le n\},$$

on $\mathbb{R}^n$. The closed (n-dimensional) cube with center x and side length $2c$ is then given by $C(x,c) = \{y : \|x - y\| \le c\}$. If $A = [a_{ij}]$ is an $n \times n$ matrix, we define the norm of A to be

$$\|A\| = \max\left\{\sum_{j=1}^{n} |a_{ij}| : 1 \le i \le n\right\}.$$

If $x \in \mathbb{R}^n$, we then have the inequality $\|Ax\| \le \|A\|\|x\|$.

Let $x \in C$ and let $C = C(x,c)$ be a cube inside D_1. By the Mean Value Theorem

$$h_i(y) - h_i(x) = \sum_{j=1}^{n} \partial_j h_i(x + \theta_j(y)(y - x))(y_j - x_j)$$

where $0 \le \theta_j(y) \le 1$. Thus,

$$\|h(y) - h(x)\| \le c \max_{z \in C} \|Dh(z)\| \text{ for } y \in C.$$

It follows that for every cube C inside D_1, we have

$$(4) \quad \lambda(h(C)) \le \left(\max_{z \in C} \|Dh(z)\|\right)^n \lambda(C).$$

Lemma 4. *If $A : \mathbb{R}^n \to \mathbb{R}^n$ is linear and C is any cube inside D_1, then*

$$(5) \quad \lambda(h(C)) \le |\det A| \left(\max_{z \in C} \|A^{-1}Dh(z)\|\right)^n \lambda(C).$$

Proof: Apply (4) to the map $A^{-1}h$ to obtain

$$\lambda(A^{-1}h(C)) \le \left(\max_{z \in C} \|A^{-1}Dh(z)\|\right)^n \lambda(C).$$

But, $h(C)$ is a bounded Borel set so (2) gives

$$(\det A^{-1})\lambda(h(C)) \le \left(\max_{z \in C} \|A^{-1}Dh(z)\|\right)^n \lambda(C),$$

and the result follows.

Lemma 5. *Given $x \in D_1$ and $\varepsilon > 0$ there exists $\delta > 0$ such that*

$$\lambda(h(Q)) \le |\det Dh(x)|(1 + \varepsilon)\lambda(Q),$$

where Q is any cube inside $C(x,\delta) \subset D_1$.

Proof: Put $A = Dh(x)$ in (5) to obtain

$$(6)\ \lambda(h(C)) \le |\det Dh(x)| \left(\max_{z\in C} \|Dh(x)^{-1}Dh(z)\| \right)^n \lambda(C)$$

where C is any cube inside D_1. Now $z \to \|Dh(x)^{-1}Dh(z)\|$ is continuous on D_1 and approaches $1 = \|I\|$ as $z \to x$ so the inequality in (6) gives the result.

We next require a variant of the Covering Lemma 6.3 for open sets in $\mathbb{R}^n$. We say that a cube in $\mathbb{R}^n$ is left-closed if it has the form $[a,b) \times \cdots \times [a,b)$ (n factors). The proof of Lemma 6.3 is easily adapted in $\mathbb{R}^n$ to give the following covering lemma (see also [M] 13.4, [MS2] 5.3, [ML] § 5.5).

Lemma 6. (*Covering Lemma*) *Let $G \subset \mathbb{R}^n$ be open and contained in a rectangle I. Let γ be a gauge on I. There exists an at most countable, pairwise disjoint family $\{Q_k : k \in M\}$ of left-closed cubes and $\{x_k : k \in M\} \subset G$ such that $x_k \in Q_k \subset \gamma(x_k)$ and $G = U\{Q_k : k \in M\}$.*

Lemma 7. *Let D_1 and D_2 be bounded and $f : D_2 \to [0,\infty)$ be continuous and bounded along with Dh and Dg. Then*

$$(7)\ \int_{D_2} f(y)dy = \int_{D_1} f(h(x))|\det Dh(x)|dx.$$

Proof: Both integrals in (7) exist by Exercise 2. Let $\varepsilon > 0$. For each $x \in D_1$ let δ be as in Lemma 5 and define a gauge γ_1 on D_1 by

$$\gamma_1(x) = \{z \in D_1 : \|x - z\| < \delta\}.$$

For each $x \in D_1$ there exists an open cube $\gamma_2(x)$ centered at x and contained in D_1 such that $|f(h(y) - f(h(x))| < \varepsilon$ and

$$|f(h(y))|\det Dh(y)| - f(h(x))|\det Dh(x)|| < \varepsilon$$

for $y \in \gamma_2(x)$. Define a gauge γ on D_1 by $\gamma(x) = \gamma_1(x) \cap \gamma_2(x)$. By the Covering Lemma there exist pairwise disjoint half-closed cubes $\{Q_k\}$ and $\{x_k\} \subset D_1$ such that $x_k \in Q_k \subset \gamma(x_k) \subset D_1$ and $D_1 = \cup Q_k$, $D_2 = \cup h(Q_k)$. If $y \in h(Q_k)$, then $y = h(x)$ for some $x \in Q_k$ so $f(y) < f(h(x_k)) + \varepsilon$, and if $x \in Q_k$, then

$$\varepsilon + f(h(x))|\det Dh(x)| > f(h(x_k))|\det Dh(x_k)|.$$

Thus, by Lemma 5,

$$\begin{aligned}\int_{h(Q_k)} f(y)dy &\le \int_{h(Q_k)} [f(h(x_k)) + \varepsilon]dy = [f(h(x_k) + \varepsilon)]\lambda(h(Q_k)) \\ &\le f(h(x_k))(1+\varepsilon)|\det Dh(x_k)|\lambda(Q_k) + \varepsilon\lambda(h(Q_k)) \\ &= (1+\varepsilon)\int_{Q_k} f(h(x_k))|\det Dh(x_k)|dx + \varepsilon\lambda(h(Q_k)) \\ &\le (1+\varepsilon)\int_{Q_k} [f(h(x))|\det Dh(x)| + \varepsilon]dx + \varepsilon\lambda(h(Q_k)) \\ &= (1+\varepsilon)\int_{Q_k} f(h(x))|\det Dh(x)|dx + \varepsilon(1+\varepsilon)\lambda(Q_k) \\ &\quad + \varepsilon\lambda(h(Q_k))\,.\end{aligned}$$

Using the Monotone Convergence Theorem and the countable additivity of λ, adding the inequality above gives

$$\begin{aligned}\int_{D_2} f(y)dy \le (1+\varepsilon)\int_{D_1} & f(h(x))|\det Dh(x)|dx \\ & + \varepsilon(1+\varepsilon)\lambda(D_1) + \varepsilon\lambda(D_2)\,.\end{aligned}$$

Letting $\varepsilon \to 0$ gives

$$(8) \quad \int_{D_2} f(y)dy \le \int_{D_1} f(h(x))|\det Dh(x)|dx.$$

In (8) we may interchange D_1 with D_2 and h with g and replace f with $f \circ h|\det Dh|$ in the left hand side of (8). Then the integral on the right hand side of (8) becomes f and gives

$$\int_{D_1} f(h(x))|\det Dh(x)|dx \le \int_{D_1} f(y)dy$$

so the result follows.

Theorem 8. *Let $f : D_2 \to \mathbb{R}$ be continuous. Then f is integrable over D_2 if and only if $f \circ h|\det Dh|$ is integrable over D_1 and (7) holds.*

Proof: First assume $f \ge 0$. For each k put

$$G_k = \{x \in D_1 : \|x\| < k,\ f(x) < k,\ |\det Dh(x)| < k,\ |\det Dg(x)| < k\}\,.$$

Then each G_k is open, bounded and increases to D_1. Lemma 7 applies to each G_k and gives

$$\int_{h(G_k)} f(y)dy = \int_{G_k} f(h(x))|\det Dh(x)|dx\,.$$

The Monotone Convergence Theorem now gives the conclusion.

If $f : D_2 \to \mathbb{R}$, the part above applies to each f^+ and f^- and gives the result.

We now establish a more general form of the transformation formula (7) for Lebesgue integrable functions. This will require fairly extensive use of results from Lebesgue measure/integration for which we will supply appropriate references.

We establish a preliminary lemma.

Lemma 9. *If $Z \subset D_1$ is (Lebesgue) null, then $h(Z)$ is null.*

Proof: Suppose that Z is bounded and is contained in the closed cube I. Given $\varepsilon > 0$ there exist cubes $\{Q_k\}$ inside I such that $Z \subset \cup Q_k$ and $\sum_{k=1}^{\infty} \lambda(Q_k) < \varepsilon$. $\|Dh(\cdot)\|$ is bounded on I by, say, M. By Lemma 5, $h(Z) \subset \cup h(Q_k)$ and $\lambda(h(Q_k)) \le M^n \lambda(Q_k)$. This implies $h(Z)$ is null.

If Z is unbounded, then $Z = \cup Z_k$, where $Z_k = \{x \in Z : \|x\| \le k\}$. Each Z_k is null by the part above so Z is null.

Theorem 10. *Let $f : D_2 \to \mathbb{R}$ be absolutely integrable over D_2. Then $f \circ h|\det Dh|$ is absolutely integrable over D_1 and (7) holds.*

Proof: Let $f_k : D_2 \to \mathbb{R}$ be continuous, integrable over D_2 such that $\|f_k - f\|_1 \to 0$ with $f_k \to f$ a.e. ([Sw1] 3.5.6, 3.7.1, 3.6.3). By Theorem 8,

$$\int_{D_2} f_k(y)dy = \int_{D_1} f_k(h(x))|\det Dh(x)|dx \to \int_{D_2} f(y)dy$$

and the sequence $\{f_k \circ h|\det Dh|\}$ is Cauchy in $L^1(D_1)$. The sequence therefore converges in $\| \ \|_1$ to some $\varphi \in L^1(D_1)$ ([Sw1] 3.5.1). But, $f_k \circ h|\det Dh| \to f \circ h|\det Dh|$ a.e. by Lemma 9 so $\varphi = f \circ h|\det Dh|$ a.e. and the result follows.

For the complete analogy of Theorem 8, see Exercise 3.

For other proofs of the transformation formula for the gauge integral see [M], [MS]. For the Lebesgue integral see [Sz] and the references therein.

Exercises

1. Complete the proof of Lemma 1.
2. Let $G \subset \mathbb{R}^n$ be bounded and open and $f : G \to \mathbb{R}$ be bounded and continuous. Show that f is integrable over G.
3. Suppose $f : D_2 \to \mathbb{R}$ is measurable and such that $f \circ k|\det Dh|$ is absolutely integrable over D_1. Show that f is absolutely integrable over D_2 and (7) holds.
   [Hint: Repeat the proof of the last part of the Lemma 7.]
4. If $E \subset \mathbb{R}^n$ is measurable, show that $h(E)$ is measurable.
5. Let $I = \int_0^\infty e^{-t^2}\,dt$. Show that $I = \sqrt{\pi}/2$ by using polar coordinates $(x, y) = (r\cos\theta, y\sin\theta)$ to compute I^2 as a double integral.

Bibliography

[A] A. Alexiewicz, Linear Functionals on Denjoy integrable functions, Colloq. Math. **1** (1948), 289–293.

[AM] C. Adams and A. Morse, On the Space (BV), *Trans. Amer. Math. Soc.* **42** (1937), 194–205.

[Ba1] R. Bartle, Return to the Riemann Integral, *Amer. Math. Monthly* **103** (1996), 625–632.

[Ba2] R. Bartle, A convergence theorem for generalized Riemann integrals, *Real Anal. Exchange* **20** (1994/95), 119–124.

[B] R. Bianconi, On the Convergence of Integrals of a Truncated Henstock-Kurzweil Integrable Function, *Real Anal. Exchange* **23** (1997/98), 247–250.

[Boas] R. P. Boas, A Primer of Real Functions, Carus Math. Monograph 13, *Math Assoc. Amer.*, Providence, 1960.

[Bo1] M. Botsko, Unified Treatment of Various Theorems in Elementary Analysis, *Amer. Math. Monthly* **94** (1987), 450–452.

[Bo2] M. Botsko, The Use of Full Covers in Analysis, *Amer. Math. Monthly* **96** (1989), 328–333.

[BBT] A. Bruckner, J. Bruckner and B. Thomson, Real Analysis, Prentice Hall, Upper Saddle River, N.J., 1997.

[BV] P. Bullen and R. Vyborny, Some Applications of a Theorem of Marcinkiewicz, *Canadian Math. Bull.* **34** (1991), 165–174.

[C] Chew Tuan-Seng, Another proof that every absolutely Henstock integrable function is McShane integrable, Functional Analysis and Global Analysis, Proc. Conference, Manila, 1996, 58–61.

[CWT] Chew Tuan-Seng, Wong Woon-Kwong and Tan Geok-Choo, On Absolutely Henstock Integrable Functions, *Real Anal. Exchange* **23** (1997/98), 799–804.

[Co] P. Cousin, Sur les fonctions de n variables complexes, *Acta Math.* **19** (1894).

[DSc] R. Davies and Z. Schuss, A Proof that Henstock's Integral includes Lebesgue, *J. London Math. Soc.* **2** (1970), 561–562.

[DS] J. DePree and C. Swartz, Introduction to Real Analysis, Wiley, N.Y., 1987.

[FM] J. Foran and S. Meinershagen, Some Answers to a Question of Bullen, *Real Anal. Exchange* **13** (1987/88), 265–277.

[GLZ] I. J. L. Garces, P. Y. Lee and D. Zhao, Moore-Smith Limits and the Henstock Integral, *Real Anal. Exchange* **24** (1998/99), 447–456.

[Gi] A. Gilioli, Natural ultrabornological, non-complete, normed function spaces, *Arch. Math.* **61** (1993), 465–477.

[Go1] R. Gordon, The Integrals of Lebesgue, Denjoy, Perron and Henstock, *Amer. Math. Soc.,* Providence, 1994.

[Go2] R. Gordon, Another Look at a Convergence Theorem for the Henstock Integral, *Real Anal. Exchange* **15** (1989/90), 724–728.

[Go3] R. Gordon, An Iterated Limits Theorem Applied to the Henstock Integral, *Real Anal. Exchange* **21** (1995/96), 774–781.

[Go4] R. Gordon, Another Proof of Measurability (of δ) for the Generalized Riemann Integral, *Real Anal. Exchange* **15** (1989/90), 386–389.

[Go5] R. Gordon, The Use of Tagged Partitions in Elementary Real Analysis, *Amer. Math. Monthly* **105** (1998), 107–117, 886.

[Go6] R. Gordon, Some Comments on the McShane and Henstock Integrals, *Real Anal. Exchange* **23** (1997/98), 329–341.

[H1] R. Henstock., Definitions of Riemann Type of Variational Integral, *Proc. London Math. Soc.* **11** (1961), 402–418.

[H2] R. Henstock, A Riemann Type Integral of Lebesgue Power, *Canadian J. Math.* **20** (1968), 79–87.

[H3] R. Henstock, Lectures on the Theory of Integration, World Scientific Publ., Singapore, 1988.

[HS] E. Hewitt and K. Stromberg, Real and Abstract Analysis, Springer-Verlag, N.Y., 1965.

[Ho] K. Hoffman, Analysis in Euclidean Space, Prentice-Hall, Englewood Cliffs, 1975.

[HK] K. Hoffman and R. Kunze, Linear Algebra, Prentice-Hall, Englewood Cliffs, N.J., 1971.

[Ku] Y. Kubota, A Note on McShane's Integral, *Math. Japonica* **1** (1985), 57–62.

[K1] J. Kurzweil, Generalized Ordinary Differential Equations and Continuous Dependence on a Parameter, *Czech. Math. J.* **82** (1957), 418–449.

[K2] J. Kurzweil, Nichtabsolut Konvergente Integrale, Tuebner, Leipzig, 1980.

[LA1] J. Lamoreaux and G. Armstrong, A Necessary and Sufficient Condition for Gauge Integrability, *Real Anal. Exchange* **19** (1993-94), 254–255.

[LA2] J. Lamoreaux and G. Armstrong, The Fundamental Theorem of Calculus for Gauge Integrals, *Math. Mag.* **71** (1998), 208–212.

[LPY] Lee Peng Yee, Lenzhou Lectures on Henstock Integration, World Scientific Publ, Singapore, 1989.

[LPY1] Lee Peng Yee, Measurability and the Henstock Integral, Proc. International Math. Conference 1994, Ed: Yuen Fong *et al.*, World Scientific Publ., Singapore, 1996.

[LW] Lee Peng Yee and Wang Pujie, Every Absolutely Henstock Integrable Function is McShane Integrable, *J. Math. Study* **27** (1994), 47–51.

[LCL] T. Lee, T. Chew and P. Y. Lee, On Henstock integrability in Euclidean space, *Real Anal. Exchange* **22** (1996-97), 382–389.

[LCL1] T. Lee, T. Chew and P. Y. Lee, Characterization of Multipliers for the Double Henstock Integrals, *Bull. Austral. Math.* **54** (1996), 441–449.

[LS] J. Lewis and O. Shisha, The Generalized Riemann, Simple, Dominated and Improper Integrals, *J. Approx. Theory* **38** (1983), 192–199.

[Lin] Lin Ying-Jian, On the Equivalence of McShane and Lebesgue Integrals, *Real Anal. Exchange* **21** (1995/96), 767–770.

[L] Liu Genqian, The Measurability of δ in Henstock Integration, *Real Anal. Exchange* **13** (1987/88), 446–450.

[Liu] Liu Genqian, On Necessary Conditions for Henstock Integrability, *Real Anal. Exchange* **18** (1992/93), 522–531.

[M] J. Mawhin, Analyse, 2nd Edition, DeBoeck Universite, 1997.

[ML] R. McLeod, The Generalized Riemann Integral, Math. Assoc. Amer., Providence, 1980.

[MS1] E. J. McShane, A Unified Theory of Integration, *Amer. Math. Monthly* **80** (1973), 349–359.

[MS2] E. J. McShane, Unified Integration, Acad. Press, N.Y., 1983.

[Me] J. Mendoza, On Lebesgue Integrability of McShane Integrable Functions, *Real Anal. Exchange* **18** (1992/93), 456–458.

[MO] P. Mikusinski and K. Ostaszewski, The space of Henstock integrable functions II, Lecture Notes #1419, Springer-Verlag, Heidelberg, 1990.

[N1] I. Natanson, Theory of Functions of a Real Variable I, Ungar, N.Y., 1955.

[N2] I. Natanson, Theory of Functions of a Real Variable II, Ungar, N.Y., 1960.

[Pe] I. Pesin, Classical and Modern Integration Theories, Acad. Press, N.Y., 1970.

[Pf1] W. Pfeffer, The Riemann Approach to Integration, Cambridge University Press, Cambridge, 1993.

[Pf2] W. Pfeffer, A Note on the Generalized Riemann Integral, *Proc. Amer. Math. Soc.* **103** (1988), 1161–1166.

[RN] F. Riesz and B. Nagy, Functional Analysis, Ungar, N.Y., 1955.

[Ro] H. Royden, Real Analysis, MacMillan, N.Y., 1988.

[Ru] W. Rudin, Principles of Mathematical Analysis, McGraw-Hill, N.Y., 1964.

[Sa] S. Saks, Sur les fonctions d'intervalle, *Fund. Math.* **10** (1927), 211–224.

[Sar1] W. L. C. Sargent, On Linear Functionals in Spaces of Conditionally Integrable Functions, *Quart. J. Math.* **1** (1950), 288–298.

[Sar2] W. L. C. Sargent, On Some Theorems of Hahn, Banach and Steinhaus, *J. London Math. Soc.* **28** (1953), 438–451.

[Sar3] W. L. C. Sargent, On the Integrability of a Product, *J. London Math. Soc.* **23** (1948), 28–34.

[Sc] A. Schurle, A Function is Perron Integrable if it has Locally Small Riemann Sums, *J. Austral. Math. Soc.* **41** (1986), 224–232.

[Sch] S. Schwabik, Henstock's Condition for Convergence Theorems and Equi-integrability, *Real Anal. Exchange* **18** (1992/93), 190–205.

[Sz] J. Schwartz, The Formula for Change of Variables in a Multiple Integral, *Amer. Math. Monthly* **61** (1954), 81–85.

[Sh] P. Shanahan, Unified Proof of Several Basic Theorems of Real Analysis, *Amer. Math. Monthly* **79** (1972), 895–898.

[Shi] O. Shisha, Tests of Existence of Generalized Riemann Integrals, Probability and Related Fields, Plenum Press, N.Y., 1994, 439–441.

[Si] M. Sion, Introduction to the Methods of Real Analysis, Holt, N.Y., 1968.

[Sw1] C. Swartz, Measure, Integration and Function Spaces, World Scientific Publ., Singapore, 1994.

[Sw2] C. Swartz, An Introduction to Functional Analysis, Marcel Dekker, N.Y., 1992.

[Sw23] C. Swartz, Even More on the Fundamental Theorem of Calculus, *Proy. Revista Mat.* **12** (1993), 129–135.

[ST] C. Swartz and B. Thomson, More on the Fundamental Theorem of Calculus, *Amer. Math. Monthly* **95** (1988) 644–648.

[Th1] B. Thomson, On Full Covering Properties, *Real Anal. Exchange* **6** (1980/81), 77–93.

[Th2] B. Thomson, Spaces of Conditionally Integrable Functions, *J. London Math. Soc.* **2** (1970), 358–360.

[Tr] F. Treves, Topological Vector Spaces, Distributions and Kernels, Acad. Press, N.Y., 1967.

[V] R. Vyborny, Kurzweil-Henstock Absolute Integrable Means McShane Integrable, *Real Anal. Exchange* **20** (1994-95), 363–366.

[We] R. Winstock, Elementary Evaluations of $\int_o^\infty e^{-x^2}dx$, $\int_o^\infty \cos x^2 dx$ and $\int_o^\infty \sin x^2 dx$, *Amer. Math. Monthly* **97** (1990), 39–42.

[YL] Ye Guoju and Lee Peng Yee, A Version of Harnack Extension for the Henstock Integral, *J. Math. Study* **28** (1995), 106–108.

[Za] A.C. Zaanen, Linear Analysis, North-Holland, Amsterdam, 1956.

Index